D1269318
CHICAGO
HAROLD WASHI

Beyond Velikovsky

Beyond Velikovsky

The History of a Public Controversy

Henry H. Bauer

University of Illinois Press
Urbana and Chicago

This book is printed on acid-free paper.

Library of Congress Cataloging in Publication Data

Bauer, Henry H.
Beyond Velikovsky.

Bibliography: p.
Includes index.
1. Astronomy—History. 2. Velikovsky, Immanuel, 1895–1979.
I. Title.
QB32.B38 1984 521'.54 83-17935
ISBN 0-252-01104-X

For Tim Dinsdale
who planted the seed

Contents

Preface

My ambition in this book is not to settle the Velikovsky controversy but to provide food for thought and a basis for understanding this and similar arguments.

Controversies of like sort abound: public controversies about scientific matters. Some such arguments split the scientific community, and the public is asked to choose between two groups of experts of opposite view, each group claiming to have the authority of science on its side: regarding the safety of nuclear power plants, the feasibility of antimissile missiles, the dangers of accumulating carbon dioxide in the atmosphere—the listing could continue at great length. In other arguments the scientific community is almost monolithic in opposing the proponents of unorthodox claims, and invokes the authority of science to convince us that UFOs are but misinterpreted balloons, planets, and so forth; that Loch Ness monsters and sea serpents do not and cannot exist; that extrasensory perception is impossible; until recently, that acupuncture does not work.

Characteristically, the public is ill served by protagonists of the opposing views. Both sides seek to win acceptance of their opinions, not to inform us about the issues. And the media rarely help: they seek "news," revel in the clamor of sharp disagreement, delight in "human-interest" trivializations, and do not analyze dispassionately in context or depth. Truth is not often told, the whole truth never.

My ambition is to use the Velikovsky affair as an exemplar of such controversies. I seek to illustrate themes that are common to these arguments, so that we may be better on guard against the propaganda:

That science always has an answer, and that scientists know what that answer is;

That because science has often been wrong, it is always likely to be so;

That scientists are impersonal, objective, and open-minded;

That because scientists are not so, therefore their claims are necessarily wrong;

That simple statements imply only what they appear to say;

That the use of numbers makes for precision and truth . . .

And much else.

I first became interested in the subject of this book by reading an account [78] of the controversy Velikovsky had caused. My curiosity was aroused: at the time (early 1975) I had been a practicing scientist for about two decades, and a student of science for almost a decade before that, and had never before heard of Velikovsky. How could there have been such a fierce controversy over a scientific issue without my being aware of it? Perhaps part of the reason was that I had been in Australia, and not in the United States, during the early 1950s and 1960s. But still, I read widely, and had friends who were physicists, astronomers, geologists. The chief reason for my previous ignorance is now clear to me: though there had indeed been fierce controversy, it had not been a *scientific* controversy. Matters pertaining to science were certainly involved, but the arguments were not carried on in the professional journals with the scientific content as the sole or chief point of dispute.

My curiosity was also aroused because—as I now realize but did not then—the presentation that I read [78] is quite one-sided, and I had at that time no immediate access to other views. So, though Velikovsky's cosmic scenario did seem inherently unlikely to me, I was greatly influenced by the book's emphasis on Velikovsky's successful predictions and by its apparent demolition of the scientists' criticisms of his work. Moreover, the accounts of conspiracy and skulduggery by prominent scientists convinced me that Velikovsky had been treated very badly, and I sympathized with the underdog. I was also not prepared to believe that the issue could not be settled on purely technical grounds, and so I was led to read Velikovsky's work itself, the criticisms of it, and finally all the material used in the present analysis.

So I began this inquiry in something of a state of surprise: that I had been unaware of the whole affair; that the issue apparently had not been decisively settled on the technical grounds alone; that there could have been such a fuss, and that scientists could have reacted so violently, if Velikovsky was just a crackpot.

Now, having worked through the record and having thought about the general issues involved, I am no longer in a state of surprise. All that happened now seems explicable to me—I flatter myself that I understand it. I think that sort of change, from a state of surprise to one of calm acceptance, is in fact a demonstration that understanding has been gained. We are surprised at things that are unexpected because, at that moment, we do not understand how they could come about.

As I came to understand the Velikovsky affair, I came also to realize that many features of that controversy characterize other controversies as well, most particularly those in which there is a high level of public interest coupled with disagreement over matters of science or technology. The plan of this book, then, is to explicate the Velikovsky affair and to indicate how an understanding of it can be applied to many other situations.

In Part I, I attempt to tell, without bias and in chronological sequence, the story of the controversy. Part II details my personal analysis of the affair and touches on wider issues that are relevant—how to distinguish cranks from scientists, for example. I attempt to make the difficult point that one can reach personal conviction without losing sight of the fact that personal convictions are not necessarily true. In Part III, I review the affair in the light of my conclusions, seeking to show that the main themes of this controversy will be found in many other public arguments, especially that gross misunderstandings about "science" abound. And I attempt a valid but nontechnical description of what science really is.

I began this work simply because I found the Velikovsky affair a fascinating controversy about which I wanted to become clear. It is my hope now that my readers will find the book not only an interesting account of that controversy but also a means of stimulating recognition of the many inaccuracies, half-truths, unsupported assertions, and misleading statements to which we all are constantly exposed.

This book has been long in the making, and I owe gratitude to many. For important encouragement and support, I thank Myra; Barbara, Paul, and Reid; and Beverly, Brauch, Don, Dorothy, Grant, and the other members of our group of the Association for Rational Thinking. Helpful comments on an early draft came particularly from Brad Canon. Alfred de Grazia was exceedingly objective and gracious in recognizing merits in a work whose nuances he himself would hardly emphasize. Marcello Truzzi helped me further to recognize where my bias had influenced me more strongly than I had wished. When I blanched at the prospect of yet another thoroughgoing rewriting, my energy was sustained by favorable comments on parts of the manuscript from people whose opinions I respect a great deal: I. J. (Jack) Good, Larry Laudan, Gordon Tullock, and John Wilson in particular. The energy and professional experience of Ann Heidbreder Eastman were of inestimable help in placing the manuscript with the right publisher. Leroy Ellenberger, who saw the manuscript as it was going to press, made it possible for me to correct some significant mistakes.

I owe a special debt of gratitude to W. T. Black, who was outstandingly generous in making available from his collection material that was otherwise difficult or impossible for me to obtain. Ellen Baxter and Jane Lane moved mountains in locating and making available many items that would have remained inaccessible to me without their help; their interest and professional assistance were invaluable. And what prodigiously rapid and endlessly patient typing and retyping Becky Cox provided. I was helped immeasurably by Carla Knapp in the verification of quotations and references in the last, crucial draft of the manuscript.

I might not have been able to bring this work to its final form but for the concern and professional competence of three physicians: William T. Hendricks of Blacksburg, J. Hayden Hollingsworth of Roanoke, and Kenneth M. Kent of the National Institutes of Health (later at Georgetown University). My gratitude to them is inexpressibly great.

Blacksburg, Virginia Henry Bauer
Summer, 1982

PART I

The Story of the Velikovsky Affair

Sir, I have found you an argument, but I am not obliged to find you an understanding. . . .

—Samuel Johnson

1

Early Publicity

There ariseth a little cloud out of the sea,
like a man's hand.
—1 Kings

Previews and Summaries

In 1946 John J. O'Neill gave readers of the *New York Herald Tribune* [278] a preview of the Velikovsky affair:

> . . . Dr. Immanuel Velikovsky, an international scholar . . . has assembled into a monumental work evidence from all the early civilizations that in the first and second millennium before Christ tremendous terrestrial cataclysms took place.
>
> In a magnificent piece of scholarly historical research he has correlated Sumerian, Chaldean, Hindu, Chinese, Mayan, Aztec, Islandic, Egyptian and Hebrew records showing that the times of cataclysms described in all of them correspond. . . . The earth, on at least two occasions, was shaken to such an extent that the prevailing calendar was thrown out of gear on its yearly basis and by several days on isolated occasions, as well as tilting its axis so that the latitude of places was changed . . . producing extensive climatic changes. . . .
>
> Dr. Velikovsky finds evidence of new planets appearing in the sky and for the earth being struck by . . . the tails of comets, and also for the existence of . . . a giant planet with a fifteen-year period with which a fifteen-year cycle in earthquakes and other calamitous events was associated.
>
> Some different interpretations undoubtedly will be assigned . . . by astronomers and physicists. . . . Dr. Velikovsky's

work, as yet unpublished, presents a stupendous panorama of terrestrial and human history which will stand as a challenge to scientists to frame a realistic picture of the cosmos.

Nothing more was heard for several years. Then, in 1950, the impending publication of Velikovsky's work was given wide publicity [324]:

> "Worlds in Collision," which Macmillan will publish on April 3, is scheduled for an unusual amount of publicity in national magazines. An article based on the book has already run in *Harper's* . . . and this was followed by articles in *Newsweek* and *Pathfinder* and an editorial in the Portland *Oregonian.* The current issue of *Collier's* contains an article called "The Heavens Burst," written by John Lear and adapted from the book. *Collier's* ran full-page advertisements in the New York *Times* and *Herald Tribune* on February 19, the day the issue appeared on the stands. An article illustrated by Chesley Bonestell will run in the March 26 issue of *This Week,* and a picture of the author, Immanuel Velikovsky, will appear with an article in the April 15 issue of *Vogue.* Fulton Oursler has written an article based on the book for the March *Reader's Digest.* . . . Macmillan will, of course, devote a nation-wide advertising campaign to the book, and there will be a self-mailing circular. . . . Sometime around publication date Dr. Velikovsky will make a "Voice of America" broadcast in German in which he will discuss his theories and his book.

The cited magazines described Velikovsky's book in glowing terms [141, 209, 217, 218, 271, 284]: a major synthesis of many disciplines, reflecting a thorough knowledge of such fields as anthropology, archaeology, astronomy, biology, chemistry, classical literature, folklore, geology, paleontology, physics, psychology, religion, world history; massive documentation from many texts—Old Testament, Talmud, Egyptian papyri—and from diverse traditions and legends: of Arabia, Babylonia, China, Finland, Greece, Iceland, India, Japan, Mexico, the Pacific Islands, Persia, Peru, Rome, Siberia, Tibet, West Africa. More than ten years of research had gone into the book [217, 218], and its conclusions were unique, revolutionary, extraordinary—contradicting prevailing views in evolution, history, and physical science, and arriving at ideas about gravitation just put forward also by Albert Einstein [271]. Velikovsky's work was comparable to the efforts of Darwin and Jeans [284] and provided support for true believers in the Old Testament [271, 284]. Since Velikovsky's views ran counter to such established ones as those of Darwin on evolution and of Newton on celestial mechanics, they were bound to be controversial and unacceptable to many scientists—but conventional scholars are fallible too, as witness their refusal, until the nineteenth century, to admit the existence of meteorites [284]. Admiration of Velikovsky's efforts was expressed by prominent individuals: John O'Neill, science editor of the *New York Herald Tribune;* Horace Kallen, former dean of the New School for Social Research; Gordon Atwater, curator of the Hayden Planetarium; the literary critic Clifton Fadiman; the Reverend Norman Vincent Peale.

Velikovsky was described as an editor, historian, and physician, with an incredible range of competence in the sciences. Born in 1895, he had been a "universal student" of natural sciences, economics, law, history, medicine, and psychoanalysis, in such diverse places as Edinburgh, Charcow, Moscow, Vienna, and Zurich. He had edited *Scripta Universitatis,* out of which grew the University of Jerusalem [209]. He had practiced medicine and psychoanalysis in Palestine, and came to New York in 1939 to write a book on Freud. Research for that work led Velikovsky to his new ideas; by 1940 the main outlines of his work were clear, and they were incorporated now in *Worlds in Collision* and two projected further volumes *(Ages in Chaos).* Although Velikovsky had not followed the accepted mode of thought of astronomers and historians, his work was a scientific correlation [217, 218], buttressed with impressive scholarship, carried through with humility and logic [284].

First Criticisms

Some scientists made immediate and strongly negative responses to those favorable opinions. David Delo of the American Geological Institute said that Velikovsky had bypassed the sound, scientific observations made by geologists during the previous century. Harlow Shapley, director of the Harvard Observatory, dismissed Velikovsky's ideas as rubbish and nonsense. Carl Kraeling, historian, director of the Oriental Institute at the University of Chicago, said Velikovsky's procedure was "apologetic"—accepting a statement as true and then looking for evidence, the reverse of the traditional scientific approach; Velikovsky was ignorant about oriental imagination and poetry, and historians could do nothing other about his work than smile and go about their business. Henry Field, anthropologist and archaeologist, and Nelson Glueck of Hebrew Union College both rejected Velikovsky's use and interpretation of biblical material [365].

Time [464] ridiculed the "Universal Scholar" who was "correcting the human race's forgetfulness" with a theory spun "out of threads snipped out of the ancient tangle of folklore." It credited Macmillan, the publishers, with "skilled use of an up-to-date technique: getting widest publicity for a doubtful article before critics have been allowed to see it. . . . In spite of the high-sounding advance testimonials . . . few experts cared to waste much time on the . . . theory. . . ."

The first attempt at detailed rebuttal of Velikovsky's ideas also appeared even before *Worlds in Collision* had been published: in *The Reporter* [287], an article by Cecilia Payne-Gaposchkin, Phillips Astronomer at Harvard. She had read with "amazement, consternation, incredulity, and derision" the piece in *Harper's* [209], and she criticized the author's ignorance of the differences between electricity and magnetism and between inertia and momentum. Velikovsky was asking that we discard "theories that can (with immense precision) interpret almost *all* known physical phenomena, in favor of a vague statement capable of explaining, not all these things, but just *one* thing—and that an event which quite possibly never occurred." Velikovsky's scenario was impossible in light of the laws of mechanics; his suggestions about magnetic effects were impossible in view of the smallness of the magnetic fields of the sun and earth; historical evidence back to 1062 B.C. contradicted the suggestion that the length of the day had changed appreciably. Velikovsky was wrong about the chemical constitution of comets, and wrong in drawing an analogy between the structure of the solar system and that of an atom. Contrary to Velikovsky, the planet Venus was known to Babylonian astronomers as early as the third millennium B.C. Velikovsky had misquoted the Bible, ascribed to Hesiod a passage from Ovid, and erred in characterizing scientists as "guarding the sacred 'Laws of Nature' against assault. . . . We welcome information that compels modification of the 'laws' we know; we spend our time trying to unearth such information. Nobody is more delighted than the man of science when discordant facts are brought to light." Scientists *were* disturbed by Velikovsky and the *Harper's* article:

> Is this scientific age so uncritical, so ignorant of the nature of evidence, that any considerable number of people will be fooled by a sloppy parade of the jargon of a dozen fields of learning? Evidently a great national magazine, and a publisher who has in the past handled great works of science, believe that they will. . . . The road to fame and fortune for the twentieth-century scholar is clear. Never mind logic; never mind the precise meaning of words or the results of exact research. Employ the vocabulary of a dozen fields of learning. Use a liberal sprinkling of Biblical phrases. . . .

The Battle Begins

Larrabee [209], Lear [217, 218], Oursler [284], and others had been impressed by Velikovsky's scholarly competence and the wide-ranging impact of his ideas on many intellectual disciplines. The tone of their articles was not dispassionate: they were announcing a major discovery rather than an as-yet-untested hypothesis; references to Darwin, Newton, and Einstein placed Velikovsky squarely among the greatest scientists of all time.

Even if Velikovsky's ideas had been in some measure compatible with existing scientific views, this publicity would have been distasteful to the academic community. In that community it is regarded as gauche, self-serving, unprofessional, simply wrong to announce discoveries in the popular press, with fanfare. One is supposed to read a paper at a professional congress or to publish in an academic journal. If the matter is of great public interest, one may (perhaps!) be forgiven for later giving information—on request—to the news media. Here, however, was a "discovery" that seemed to the professional scientist to be spurious, mistaken, without redeeming value, and publicized most unprofessionally. It is not surprising, then, that Payne-Gaposchkin's tone also was not dispassionate; she did not merely say, here is why Velikovsky's ideas are unacceptable as far as astronomy is concerned—in tones of derision she answered the tones of acclamation and adulation used by Larrabee and the others.

Inevitably, then, the battle was joined, and became more heated. To disentangle the arguments, one would have to wade in polemic and counterpolemic, with no quarter given on either side. Nor would one find either side prepared to admit to error even in minor matters.

Thus "Larrabee Rebuts" in the correspondence column of *The Reporter* [210]: Payne-Gaposchkin's statements of fact were irrelevant or inexact (with one exception), presumably because she had not read the book she was criticizing. Velikovsky admitted and discussed his differences with Newtonian theory. Payne-Gaposchkin was wrong about early Babylonian observations of Venus, about the biblical quotations, and about the provenance of the passage from Hesiod.

And, in reply, "Payne-Gaposchkin Stands" [288]:

> . . . I have obtained an advance copy of *Worlds in Collision* . . . [it is] better written and more fully documented than the popularizing previews, but . . . just as wrong.
>
> I *have* read Dr. Velikovsky's "discussion of gravitational theory" which he set forth several years ago in a pamphlet entitled "Cosmos without Gravitation." I find it completely unconvincing; it could be answered point by point, but the answer would necessarily be long and technical.
>
> . . . I selected one *general* statement of Sarton that the [Babylonian] observations are of early date, and one *specific*

> and convincing early dating from Langdon and Fotheringham. There is nothing inconsistent in these two statements. . . .
>
> The Biblical accounts . . . do not mention a "blast." . . . A "blast" *is* mentioned in a *prophecy* made by Isaiah *before* the event. I am informed by a Hebrew scholar that the original word means "wind" or "spirit," certainly not "fire."
>
> The story . . . does not occur in the *Theogony,* or anywhere else in Hesiod. A Phaethon is mentioned in Hesiod, but not in connection with any such story.

Others joined in the fray. Herbert B. Nichols, science editor of the *Christian Science Monitor,* derided "The Velikovsky Excursion" [274]:

> . . . Not since Capt. Heinie Hasenpfeffer was reported sailing into New York harbor with a cargo of subways and artesian wells has there been a better candidate for P. T. Barnum's Hall of Fame. . . .
>
> The silence of scholars on Velikovsky would be appalling in the face of such flimsy evidence if it were not for the high standard of living we enjoy, which offers daily proof that the basis of modern technology cannot be very far wrong in its own material sphere of activity. . . .
>
> The role of critic of exciting theories born of wishful thinking is arduous, time consuming, and thankless. It seems some people believe what they want to believe regardless of logic. But if any reputable scientist comes forth publicly to back Velikovsky—I for one promise to join anybody who wants to stake out real estate on the moon, build a perpetual motion machine, or equip a safari to search for the sidehill wampus, tripodero, Lochness Monster, or the whirling whimpus.

From the very beginning, then, opinions about Velikovsky's work were strongly polarized, highly favorable or totally negative. One of the very few who tried to find middle ground was Gordon Atwater [18]. He expressed disagreement with many of Velikovsky's propositions but acclaimed the attempt at an unusual and interdisciplinary approach: a time would come when the initial hostility of scientists would give way to the recognition that something of value might be found in the mass of evidence and ideas assembled by Velikovsky.

Velikovsky himself was heard from publicly at this time, in the Letters section of *Harper's* [426]: "*To the Editors:* I believe that fairness requires that no judgment be pronounced on my forthcoming book by people who have not read it. If, however, there are scholars who analyze and reject *Worlds in Collision* without having seen it, then please do not refuse to print their letters, so that everyone may know who they are. Then they will stand before the public as scholars who write before they read."

Public Reactions

Tremendous public interest had already been aroused. The issue of *Harper's* that contained Larrabee's article soon sold out, and some daily newspapers reprinted the article in full [172]. The correspondence columns of the magazines and newspapers reflected the wide spread of opinion: there were letters pro, and letters con, and other letters too. In the *Christian Science Monitor* Nichols's piece was criticized for its tone of ridicule; for the claim that our high standard of living proves that "the basis of modern technology cannot be very far wrong" [457]; and for characterizing a Christian belief in miracles as embracing outrageous improbability [469]. *Time* was lauded [194] for counteracting the articles favorable to Velikovsky's work and castigated [224] for its "singularly biased stand" in doing that. *Harper's* was informed that it had been hoodwinked into presenting fancy as fact, pseudo-science as science, thereby "insult[ing] every sound scientist" and making "all the fact articles in all issues of *Harper's* . . . suspect" [215, 226, 246, 267, 382, 383].

2

Worlds in Collision

> Search well another world; who studies this,
> Travels in clouds, seeks manna, where none is.
> —Henry Vaughan

The formal publication date for the book that had already led to such controversy was 3 April 1950. "The first three printings were sold out before publication . . . and within a week of publication there were 55,000 copies in print . . ." [325]. In May the book "shot into first place among best sellers . . . among general non-fiction, [it] . . . was being outsold by only one book—the Bible" [83].

A not inconsiderable number of people have accepted as valid much of Velikovsky's work. *Worlds in Collision* has been reprinted many times (some 72 times in English alone by 1974 [445]), and later books by Velikovsky have also enjoyed good sales in hardcover and paperback editions.[1] The *American Behavioral Scientist* devoted an entire issue [8] to the Velikovsky controversy; an augmented version of that issue was later published as a book [78]. Periodicals largely concerned with Velikovskian studies have appeared—*Pensée* [291], *Kronos* [199], *S.I.S. Review* [346]; some of the material from *Pensée* has been published in book form [315]. Sym-

1. A comparison may be useful. Leroy Ellenberger has (1983) traced the evidence on the issue of how impressive a best-seller *Worlds in Collision* was. For the whole of 1950, it was not in the top ten though not far behind. *Kon Tiki* (published only in September) was fifth with sales of more than 128,000; *Dianetics* ranked twelfth with about 80,000; *Worlds in Collision* probably sold about 75,000 that year.

posia have been held to discuss Velikovsky's ideas, a recent book is entitled *The Age of Velikovsky* [331], and part of the proceedings of one symposium has been published as *Scientists Confront Velikovsky* [119].

One can learn a great deal from these various secondary sources, but if one wishes to reach a genuinely informed opinion, it is also necessary to read for oneself what Velikovsky actually wrote. His books cannot be summarized in a completely unbiased fashion; one selects what seems most significant, and opinions about what that is will inevitably differ. I urge the reader to go to Velikovsky's books and to judge them for himself. Velikovsky was very sensitive about attempts to summarize his works, and perhaps he had good reason to be. For the present purpose, however, I must say something about *Worlds in Collision;* the reader is reminded that the following selection is made for a specific purpose and does not pretend to do full justice to the tone or substance of Velikovsky's book [408]. But I do attempt to reflect those attributes without deliberate bias.

The preface to *Worlds in Collision* illuminates Velikovsky's approach and attitude:

> *Worlds in Collision* is a book of wars in the celestial sphere . . . the planet earth participated . . . one . . . occurred . . . in the middle of the second millennium before the present era; the other in the eighth and the beginning of the seventh century before the present era. . . .
>
> Harmony or stability in the celestial and terrestrial spheres is the point of departure of the present-day concept of the world as expressed in the celestial mechanics of Newton and the theory of evolution of Darwin. If these two men of science are sacrosanct, this book is a heresy. However, modern physics . . . describes dramatic changes in the microcosm—the atom—the prototype of the solar system; a theory . . . that envisages not dissimilar events in the macrocosm—the solar system—brings the modern concepts of physics to the celestial sphere.
>
> This book is written for the instructed and uninstructed alike. No formula and no hieroglyphic will stand in the way of those who set out to read it. If, occasionally, historical evidence

> does not square with formulated laws, it should be remembered that a law is but a deduction from experience and experiment, and therefore laws must conform with historical facts, not facts with laws.
>
> The reader . . . is invited to consider for himself whether he is reading . . . fiction or non-fiction, . . . invention or historical fact. . . .
>
> It was in the spring of 1940 that I came upon the idea that in the days of the Exodus, as evident from many passages of the Scriptures, there occurred a great physical catastrophe, and that such an event could serve in determining the time of the Exodus in Egyptian history. . . . Already in the fall of that same year, 1940, I felt that I had acquired an understanding of the real nature and extent of that catastrophe, and for nine years I worked on [*Worlds in Collision* and *Ages in Chaos*]. . . .
>
> If cosmic upheavals occurred in the historical past, why does not the human race remember them . . .? I discuss this . . . in . . . "The Collective Amnesia." The task I had to accomplish was not unlike that faced by a psychoanalyst who, out of disassociated memories and dreams, reconstructs a forgotten traumatic experience in the early life of an individual. In an analytical experiment on mankind, historical inscriptions and legendary motifs often play the same role as recollections (infantile memories) and dreams in the analysis of a personality.

In a prologue following the preface, Velikovsky summarizes facts and theories about the solar system and points to many questions that remain unanswered: about the origins of the planets, of comets, of life, of the ubiquitous legend of the Flood; about the causes of mountain building and of the Ice Ages, for example. The conflict is stressed between modern theories of gradual evolution (Lyell, Darwin) and the earlier one of catastrophic change (Cuvier). Velikovsky shows that the concept of ages brought to an end by violent natural changes can be found in the traditions of peoples the world over, in Armenia, China, Etruria, Greece, Iceland, India, Mexico, Persia, Polynesia, and Tibet, among others.

Velikovsky's main argument begins with the biblical story that the sun stood still at Joshua's command: "A departure of the earth from its regular rotation is thinkable, but only in the very improbable event that our planet should meet another heavenly body of sufficient mass. . . . That a comet may strike our planet is not very probable, but the idea is not absurd. . . . If the head of a comet should pass very close to our path, . . . another phenomenon besides the disturbed movement of the planet would probably occur: a rain of meteorites would strike the earth . . ." [408:40–44]. Both a change in the apparent path of the sun—i.e. a change in the earth's rotation—and the fall of stones from heaven are recorded in the Book of Joshua. "The author of the Book of Joshua was surely ignorant of any connection between the two phenomena. . . . As these . . . were recorded to have occurred together, it is improbable that the records were invented" [408:40–44]. Moreover, traditions from the Western Hemisphere tell of a long night, which would correspond with Joshua's long day in the Near East. Therefore, Velikovsky concludes, there actually was an apparent change in the sun's regular motion, accompanied by the fall of meteorites, and the simultaneous occurrence of these two phenomena is best explained by an encounter between the earth and a comet. But a sudden stopping of the earth's rotation would lead to total destruction, so presumably some mechanism slowed the process, or perhaps the earth's axis was tilted without disturbing its rotation [408:40–44].

Further, the Central American traditions tell of a catastrophe that preceded, by 52 years, the long night; and 52 years before Joshua's miracle is a likely time for the Exodus. For this earlier catastrophe, marked by the "plagues of Egypt," Velikovsky quotes tales from all over the globe of apparently similar events: rivers turning to blood (Mayan, Egyptian, biblical, Greek, Finnish stories); "the descent of a sticky fluid which came earthward and blazed with heavy smoke is recalled in the oral and written traditions of the inhabitants of both hemispheres" [408:54] (Mayan, Siberian, East Indian, Egyptian, biblical, Babylonian); and fires, earthquakes, and the like. These events, and a long period of darkness, Velikovsky ascribes to an earlier encounter with the comet, from which fell red dust, meteorites, petroleum, and other substances. He envisages also tremendous electrical discharges and spectacularly visible variations in position and shape of the comet. These latter, Velikovsky believes, are reflected in historical and mythological accounts of events in the skies and of battles between gods (the planets).

Finally, Velikovsky identifies the comet as what is now the planet Venus. Venus, he says, originated by expulsion or fission from Jupiter and roamed the solar system as a comet, experiencing close encounters with the earth at 52-year intervals in the middle of the second millennium B.C. The comet then changed the orbit of Mars, and the latter had close encounters with earth in the seventh and eighth centuries B.C.; meanwhile, the comet had been pulled by the sun into a planetary orbit. In support, Velikovsky cites references to Venus that seem incongruous as descriptions of a planet but plausible as descriptions of a comet: "The early traditions . . . of Mexico . . . relate that Venus smoked"; "it is . . . said in the *Vedas* that . . . Venus looks like fire with smoke. . . . in the Talmud, . . . 'Fire is hanging down from the planet Venus'" [408:163–64]; and descriptions from various peoples of Venus having a beard or horns. Thus Velikovsky builds his case, with innumerable quotations from records of many ages and many parts of the world. His book contains more than 300 pages of such correlations and deductions.

Advance Claims

In the last chapter of *Worlds in Collision* (as well as in some earlier sections of the book) Velikovsky states his conclusions about some of the characteristics of the moon, Venus, and Mars—their thermal balance, the gases in their atmospheres, and so on. Little or no direct observational data about these characteristics were available at the time Velikovsky wrote, and one could therefore regard these statements as predictions of what would be found; Velikovsky himself preferred to call these statements "advance claims" rather than "predictions." Since the accuracy (or otherwise) of these advance claims was much discussed in the ensuing controversy, they are reported here in some detail.

Craters of the Moon

There are two widely accepted explanations, not mutually exclusive, for the origin of the lunar craters: volcanic activity or the impact of meteorites. Not so, according to Velikovsky:

> the moon . . . passed through the fabric of the great comet at the time of the Exodus, and in . . . the eighth century before the present era, the moon was more than once displaced from its orbit by Mars. During these catastrophes the moon's surface flowed with lava and bubbled into great circular formations. . . . In these . . . collisions or near contacts the surface . . . was also marked with clefts and rifts. . . .
>
> The great formations of craters, mountains, rifts, and plains of lava . . . were formed not only in the upheavals described in this book, but also in those which took place in earlier times. [408:361–64]

Life on Mars

> The contacts of Mars with other planets larger than itself and more powerful make it highly improbable that any higher forms of life, if they previously existed there, survived on Mars. . . . [408:361–64]

The Atmosphere of Mars

> . . . In . . . contacts between Venus, the earth, and Mars there was an exchange of atmospheres. . . . The white precipitated masses on Mars, which form the polar caps, are probably of the nature of carbon, having been acquired from the trailing part of Venus, and only the difference in atmospheric conditions on Mars as compared with the earth, together with a difference in temperature, keeps this "manna" from being permanently dissolved under the rays of the sun.
>
> The main ingredients of the atmosphere of Mars must be present in the atmosphere of the earth. . . . As oxygen and water vapor are not the main ingredients of the atmosphere of Mars, some other elements of the terrestrial atmosphere must be. . . . It could be nitrogen. . . .

. . . argon and neon are present . . . in the air. These . . . excite spectral lines only when in a hot state; . . . they cannot be detected through lines of emission from . . . Mars. The absorption lines of argon and neon have not yet been investigated. When a study of these lines will make possible a spectral search for these rare gases on planets, Mars should be submitted to the test. If analysis should reveal them in rich amounts, this would . . . answer the question: What contribution did Mars make to the earth when the two planets came into contact? [408:366–71]

The Thermal Balance of Mars

. . . Mars emits more heat than it receives from the sun. . . . What . . . is the cause of the excess of heat in Mars? . . .

The contacts of Mars with Venus, and in a lesser degree with the earth, less than three thousand years ago probably are responsible for the present temperature of Mars; interplanetary electric discharges could also initiate atomic fissions with ensuing radioactivity and emission of heat. [408:366–71]

Origin of Terrestrial Petroleum

A part of the gaseous trail of Venus remained attached to the earth, . . . some became a deposit of petroleum. . . . [408: 366–71]

Crude petroleum is composed of . . . carbon and hydrogen. The main theories of the origin of petroleum are: 1 . . . Hydrogen and carbon were brought together in the rock formations of the earth under great heat and pressure. 2 . . . Both the hydrogen and carbon . . . come from the remains of plant and animal life, in the main from microscopic marine and swamp life. . . .

The tails of comets are composed mainly of carbon and hydrogen gases. . . . If [these] . . . enter the atmosphere in huge masses, a part of them will burn . . . the rest will escape combustion . . . will become liquid . . . sink into the pores of the sand and into clefts between the rocks. . . . [408:53]

The rain of fire-water contributed to the earth's supply of petroleum; rock oil in the ground appears to be, partly at least, 'star oil' brought down at the close of world ages, notably the age that came to its end in the middle of the second millennium before the present era. [408:57]

The Gases of Venus

A part of the gaseous trail of Venus . . . followed the head of the comet . . . what remained forms today the envelope of carbon clouds of the Morning Star. . . .

If the petroleum that poured down on the earth . . . was formed by means of electrical discharges from hydrogen and gaseous carbon, Venus must still have petroleum because of the discharges . . . between the head and tail of the comet. . . . [408:366–71]

Petroleum on Jupiter

. . . If . . . Venus was thrown off from Jupiter . . . and if Venus has petroleum gases, then Jupiter must have petroleum. The fact that methane has been discovered in the atmosphere of Jupiter—the only known constituents of its atmosphere are the poisonous gases methane and ammonia—makes it rather probable that it has petroleum; the . . . "natural gas" found in and near oil fields consists largely of methane. [408:366–71]

Life on Jupiter and Venus

The modern theory of the origin of petroleum . . . regards . . . [it] as originating from organic, not inorganic, matter. Consequently, if I am not mistaken, Venus and Jupiter must possess an organic source of petroleum. . . . it was shown that there are . . . historical indications that Venus—and therefore also Jupiter—is populated by vermin; this organic life can be the source of petroleum. [408:366–71]

The Thermal Balance of Venus

What explanation can be given for the phenomenon of the nearly uniform temperature of the day and night hemispheres of Venus? The conclusion drawn was this: The daily rotation of the planet . . . is very rapid and during the short night the temperature cannot fall to any considerable extent. But this . . . stands in complete contradiction to what was believed to be the established fact of the nonrotation of Venus with respect to the sun. . . .

In reality there is no conflict between the two methods of physical observation. The night side of Venus radiates heat because Venus is hot. . . .

> Venus experienced in quick succession its birth and expulsion under violent conditions; an existence as a comet . . . which approached the sun closely; two encounters with the earth accompanied by discharges of potentials between these two bodies and with a thermal effect caused by conversion of momentum into heat; a number of contacts with Mars, and probably also with Jupiter. Since all this happened between the third and first millennium before the present era, the core of the planet Venus must still be hot. Moreover, if there is oxygen present . . . petroleum fires must be burning there. [408:366–71]

Summing Up

Velikovsky concludes that cosmic collisions are "implicit in the dynamics of the universe" [408:374]. His ideas provide explanations for such diverse matters as: the origins of a planet, of comets, of surface features of earth, moon, and Mars; the extinction of species; the building of mountains; the great migrations of peoples in the fifteenth and eighth centuries B.C.; and the origin of the belief in the chosenness of the Jewish people [408:379–81].

> The theory of cosmic catastrophism can, if required to do so, conform with the celestial mechanics of Newton. . . . [408:384]

> . . . I must admit, however, that . . . I became skeptical of the great theories concerning the celestial motions that were formulated when the historical facts described here were not known to science. . . . I would venture to say . . . [that] accepted celestial mechanics . . . stands only *if* the sun . . . *is as a whole an electrically neutral body,* and . . . if the planets . . . are neutral bodies.
>
> When physicists came upon the idea that the atom is built like a solar system . . . the notion was looked upon with much favor. But . . . "an atom differs . . . by the fact that it is not gravitation that makes the electrons go round the nucleus, but electricity" . . . another difference was found: an electron in an atom, on absorbing the energy of a photon (light), jumps to another orbit, and again to another when it emits light. . . .
>
> . . . The solar system is actually built like an atom; only, in

> keeping with the smallness of the atom, the jumping of electrons from one orbit to another, when hit by the energy of a photon, takes place many times a second, whereas in accord with the vastness of the solar system, a similar phenomenon occurs there once in hundreds or thousands of years. In the middle of the second millennium . . . the terrestrial globe experienced two displacements; and in the eighth or seventh century . . . three or four more. In the period between, Mars and Venus, and the moon also, shifted.
>
> If the activity in an atom constitutes a rule for the macrocosm, then the events described in this book were not merely accidents of celestial traffic, but normal phenomena like birth and death. The discharges between the planets, or the great photons emitted in these contacts, caused metamorphoses in inorganic and organic nature. [408:387–89]

Finally, "having discovered some historical facts and having solved a few problems, we are faced with more problems in almost all fields of science. . . . Barriers between sciences serve to create the belief in a scientist in any particular field that other scientific fields are free from problems, and he trusts himself to borrow from them without questioning. It can be seen here that problems in one area carry over into other scientific areas, thought to have no contact with each other" [408:387–89].

Reviews

Worlds in Collision was reviewed in innumerable newspapers and magazines. Favorable reviews echoed the tone of the magazine articles mentioned earlier. Velikovsky was compared with Darwin, Einstein, Galileo, Kepler, Newton, Planck; his competence over the whole field of learning was stressed. His ideas were said to challenge or upset prevailing theories in many disciplines, and to be remarkable, absolutely original and convincing, revolutionary, stupendous in implication. This "most sensational book of the 20th century," "one of the most significant and fascinating . . . written since the invention of printing," "may rock the civilized world" and "has literally shaken the thinking world to its foundations"; "it may well become a bible." "Top-flight scientists think . . . [it] will force reconsideration of the basic postulates of many major sciences." "The minds that will be stimulated . . . are the world's best" [92].

Favorable reviews were at least as common as unfavorable ones, but I shall devote more space here to the latter: those reviews that commended Velikovsky did so in the rather general terms reported above, and can be quickly summarized; the negative reviews, on the other hand, mentioned many points of detail.

Several of the most scathing reviews were written by scientists, astronomers in particular—Harrison Brown [37], Edward Condon [56], Paul Herget [146], Kirtley Mather [238], Cecilia Payne-Gaposchkin [289], Spencer Jones [163], Otto Struve [393], Rupert Wildt, K. S. Latourette, and Chester R. Longwell [214]. Some writers and literary critics also found the book unconvincing, among them W. E. Garrison [115], Ben Hunt [153], Alfred Kazin [185], and George R. Stewart [289].

One of the arguments raised against Velikovsky's approach and conclusions was his use of historical records. For example:

Impossible synchronizations and wrong datings—of Aztecs and Toltecs [214]; of Gilgamesh's epic and the Exodus [289]; of the clay tablets of Nineveh [269].

Uncritical use of accounts—about the Chinese Emperor Yao, who may not even have existed; from widely different times about events at Mt. Sinai [214]; Ovid's later work rather than the original Homer and Hesiod; rabbinical commentaries rather than the biblical narratives [289].

Biased selection—ignoring eight other possible dates for the Mexican New Year, all equally authoritative as the one used [214]; ignoring the reported fission of the moon (Koran) and the sun (Volospa) [139], and many other cosmic myths that do not fit his particular purpose; ignoring the records of early eclipses (from 762 to 719 B.C. [269], 1062 B.C. [289], 2137 B.C. [393]), which show that the earth's motion was essentially the same in those times as now; ignoring other early records showing that the apparent motions of the sun [139] and of Venus [179] have remained unaltered.

Far-fetched interpretations—about Typhon and Phaethon [289]; of biblical evidence for a 30-day month [115].

Misleading quotation—about the systems of the planets and their motions, a 360-day year, and the Chaldean ecliptic (the latter being a mistranslation from the German) [269]; about the destruction of Sennacherib's army [289].

Arbitrariness—reducing a reported date of 9,000 years ago to 900 only because that fits his own predilection [384].

Ignorance—of the reason for the Mesoamerican 52-year cycle [214].

Velikovsky was also accused of incompetence in the natural sciences. He lacks understanding of the law of gravitation [214] and does not know that comets are very much smaller than Venus [163, 214]. His analogy between the atom and the solar system is not valid [179]. Precise calculations, which fit existing facts, show that the planets did not encounter one another within the last hundreds of thousands of years [163, 214]. Roche's limit shows that close encounters would have led to disintegration of one or both of the bodies involved [179]. The structure of oil-bearing deposits precludes the possibility that the oil entered from above, and the "erratic blocks" are completely explicable in terms of glacial action [214]. Bands of mud deposits, and California tree-rings, indicate no major climatic disturbance in the last 3,000 years and no major worldwide inundation in the last 15,000 years [139].

Further, several examples were given of lack of self-consistency in Velikovsky's arguments. One concerns a year of other than 365 days [214]. Ancient star maps quoted by Velikovsky himself also contradict his point [139]. His reasoning is circular—using records to establish his hypothesis, then "confirming" the latter by reference to the records [185].

These points were supported by more or less detailed discussion. But many critics also dismissed all of Velikovsky's work in *ex cathedra* fashion: "The physical, chemical and astronomical statements . . . are all so completely at variance with known principles . . ." [56]. "It is . . . unnecessary to waste . . . space . . . pointing out the numerous errors in . . . scientific principles. . . . Anyone with only an elementary knowledge of astronomy, physics, chemistry, or geology will immediately detect them" [238]. The "sincere musings of a man . . . unfamiliar with the details and general principles of the physical sciences . . ." [37].

When a book is unfavorably reviewed, it is not uncommon for the reviewers to display derision and sarcasm. Velikovsky was not spared:

> By a merciful ordination of the economic system . . . the host of amateur cosmologists seldom command the means of bringing their speculations before the eyes and ears of the world. The author . . . has scored over his brethren. . . . his real coup had been to launch in the Scripta Academica Hierosolymitana . . . "Cosmos without Gravitation" [published privately in 1946]. . . . readers [of *Worlds in Collision*] . . . are spared the realization that its author ever professed belief in . . . "the empiric evidences of the fallacy of the law of gravitation." . . . We look in vain for an explanation of what possessed the man between 1946 and 1950 . . . is this . . . individual amnesia overtaking the author, or does he have so little respect for scientific critics as to rely on their collective amnesia? [214]

> . . . I assume that a failure to correlate the Volospa with Lear's works prevented him from adding the Fimble Fowl . . . to his synonyms for Venus. . . . I could write as convincing a book . . . to prove that monkeys had originated from men . . . the book is fiction, and I think its author has deliberately left several clues to this effect . . . statements that methane is a poisonous gas, that "the absorption lines of argon and neon have not yet been investigated," and that flies may well live on Jupiter, were probably intended to warn off chemists, physicists, and zoologists. . . . the kindest judgment . . . is to class him with Defoe and Samuel Butler II in the select company of successful hoaxers rather than in the much larger army of cranks. [139]

> The library of the Yerkes Observatory has a shelf . . . reserved for the screwball fringe of science . . . astrology, the flat earth, the flying saucers. . . . Our latest addition . . . is

> "Worlds in Collision" . . . not a book of science and it cannot be dealt with in scientific terms. [393]

And there were many other comments of like nature [185, 238, 384, 401].

Conspiracy and Skulduggery

Some of Velikovsky's detractors were apparently prepared to go further than the writing of condemnatory reviews: "The biggest bombshell dropped on Publisher's Row in many a year exploded . . . when . . . Macmillan . . . dumped . . . 'Worlds in Collision' overboard . . . [with the] formal announcement that 'by mutual agreement between the author and publisher' the top non-fiction best seller would henceforth be published by Doubleday. . . . a publishing official admitted, privately, that a flood of protests from educators and others had hit the company hard in its vulnerable underbelly—the textbook division. . . . Macmillan reluctantly succumbed. . . . Was it censorship? Or public opinion?" [84].

According to Velikovsky,

> Macmillan . . . was subjected to pressure on the part of a little and seemingly organized group of scientists. . . . he quoted George P. Brett, president . . . that "a boycott of the textbook department is being organized." . . . James Putnam, his editor at Macmillan, had resigned . . . pressure had also been exerted on the Hayden Planetarium which, under the direction of Gordon A. Atwater, had prepared a show based on the book. . . . "Mr. Atwater received an order not to proceed with the show." . . . Atwater . . . had since resigned . . . as Curator of the . . . Planetarium and chairman of the department of astronomy at the American Museum of Natural History. . . . [326]
>
> . . . Off the record, Macmillan's competitors reported that a boycott had actually been started. . . . Macmillan salesmen were finding that at several universities . . . the professors wouldn't even talk. . . . In the New York *Post* . . . Leonard Lyons declared that it was an organized campaign and the leader . . . was Dr. Harlow Shapley. . . . [272]

Shapley denied "that he conducted 'any kind of campaign against the book. . . . I didn't make any threats and I don't know anyone who did'" [272]. "The claim that Dr. Velikovsky's book is being suppressed is nothing but a publicity promotion stunt. Like having a book banned in Boston, it improves the sales. Several attempts have been made to link such a move to stop the book's publication to some organization or to the Harvard Observatory. This idea is absolutely false" [90].

There was, however, at least circumstantial evidence of Shapley's involvement. Most of the complaints addressed to the Macmillan Company came from astronomers. Shapley was president of Science Service, whose *Science News Letter* had been the first to denounce Velikovsky's book, even before its publication. The first detailed rebuttal of Velikovsky came from Payne-Gaposchkin, a colleague of Shapley at Harvard [272]. And in a letter [90] to a friend Shapley had apparently acknowledged that his efforts were in part responsible for Macmillan's decision to transfer *Worlds in Collision* to Doubleday.

This series of events—boycott of Macmillan, firing of Atwater and Putnam,[2] alleged conspiracy by prominent as-

2. In Dec. 1979 I spoke by telephone with Gordon Atwater, then 76 years old. Atwater still feels keenly aggrieved at the manner in which Velikovsky was treated. He plans to publish his own account of these events; on 12 Dec. 1979 he related his experiences to students at the New School for Social Research in New York [460]. In 1950, after five years of service, Atwater was dismissed from his position at fifteen minutes' notice and not permitted to occupy his offices even long enough to remove his personal effects—his books and papers were later sent to him. According to Atwater, he was also effectively "black-listed" and was unsuccessful in attempts thereafter to obtain a position in science education. *Who's Who in America (1950–51)* mentions that Atwater had studied engineering at Purdue University and with the U.S. Power Squadrons; he was on active duty with the U.S. Navy (1942–45) in a capacity that comprised education, management, and technical development; he was a member of the American Astronomical Society and the Explorers Club, an honorary member of the Amateur Astronomers Association, a charter member of the Institute of Navigation, and a Fellow of the Royal Astronomical Society.

No details about Putnam's dismissal or his later activities were available in the literature of the Velikovsky affair before 1984, when material based on his papers and relevant Macmillan files—found in the New York Public Library—was made public (see pp. 82, 246n6).

tronomers—brought new dimensions to the controversy about Velikovsky's ideas. His detractors—even if only a small group of them—had acted unethically, and all his critics would henceforth stand in danger of being tarred with the same brush as the presumed conspirators. College professors and scientists were castigated for themselves doing what they traditionally fought against: suppressing ideas, assaulting academic freedom, acting dogmatically and in authoritarian fashion [92E, 153, 401, 272, 362]. And this raised another question: if Velikovsky was so clearly wrong, why was this not demonstrated clearly, convincingly, once and for all? "If there is nothing to Dr. Velikovsky's thesis, why were so many people trying to discredit and silence him?" [90].

Perhaps, then, the small group of astronomers was wrong not only in the methods it used but also about Velikovsky's ideas. "Early reviewers tended to follow the line of attack set up by Shapley and Mrs. Payne-Gaposchkin; but a new note of respect began to color the response of independent readers. . . . Robert H. Pfeiffer, lecturer on Semitic languages, wrote Dr. Velikovsky, 'I was amazed at the depth and vastness of your erudition. . . .' Professor Walter S. Adams, head of the Mount Wilson Observatory, while taking exception to specific points of science, complimented the author on the research as a whole" [90].

3

The First Battle

And we are here as on a darkling plain
Swept with confused alarms of struggle and flight,
Where ignorant armies clash by night.
—Matthew Arnold

In this chronological account of the Velikovsky affair, the period up to 1962 will be treated separately from the succeeding years, for reasons that will become obvious later. The first skirmishes have already been described. Battle proper may be said to have been joined when Velikovsky replied in some detail to his critics.

Velikovsky wrote letters to *The Spectator* [427] and to the *New Statesman and Nation* [428] to answer criticisms contained in reviews, in those magazines, of *Worlds in Collision.* For *Harper's* [429] he wrote a longer article as a general reply to his critics. The same issue of *Harper's* had a commentary [385] on Velikovsky's answer written by an astronomical physicist, John Q. Stewart, followed by a rebuttal [430] from Velikovsky. He also gave a full discussion in an address to graduate students at Princeton University, and that address was later published as a supplement [431] to Velikovsky's book *Earth in Upheaval* [411].

Critics and supporters of Velikovsky exchanged words in various magazines—*Harper's, Time, Science, Scientific American*—and analyses or editorial comments appeared in a number of other places. Several major attacks on Velikovsky appeared: he was classed as a crank or pseudo-scientist in articles in *Scientific Monthly* [207] and *Antioch Review* [112]; in a symposium of the American Philosophical Society on "Some Unorthodoxies of Modern Science," later published [34, 52, 187, 290, 338]; and in Martin Gardner's book *In the Name of Science* [113].

During the period of this first battle three other books by Velikovsky were published, all by Doubleday: in 1952, *Ages in Chaos* [410]; in 1955, *Earth in Upheaval* [411]; in 1960, *Oedipus and Akhnaton* [412]. None of these books caused nearly so much fuss as had *Worlds in Collision.* Reviews were again mixed, and frequent reference was made to that earlier book and the controversy it had generated.

Ages in Chaos

Velikovsky had actually written *Ages in Chaos* and *Worlds in Collision* concurrently. Both works stemmed from identifying accounts in the Bible and in the Papyrus Ipuwer as of the same set of actual events, a cosmic catastrophe during which the Exodus took place. *Worlds in Collision* describes the physical events on earth and in the heavens, whereas *Ages in Chaos* deals with the social and political histories.

In the recorded history of Egypt as conventionally interpreted, Velikovsky says, one finds no references to events of major significance in the history of the Jews—the "glorious age of King Solomon," the time of bondage in Egypt, and the Exodus. But descriptions of natural disasters in the Papyrus Ipuwer are reminiscent of the "plagues," described in the Bible, that afflicted Egypt around the time of Exodus. Velikovsky concludes that Papyrus Ipuwer was a contemporaneous account of events during the Exodus. But then, in order to maintain synchrony between other events described in both Egyptian and Hebrew records, the conventionally accepted time scales have to be altered: "Either Egyptian history is much too long or biblical history is much too short . . ." [410:336–37].

Using various sources, interpretations, and assumptions,

Velikovsky changed the accepted synchronizations and proposed new ones—"the Israelites left Egypt on the eve of its invasion by the Hyksos"; "the Hyksos domination of Egypt was the time of the Judges in scriptural history"; "David was a contemporary of Ahmose, founder of the Eighteenth Dynasty, and of Amenhotep I; Solomon was a contemporary of Thutmose I and Hatshepsut . . . described in the Scriptures as . . . the Queen of Sheba" [410:336–37]; and so on.

That Velikovsky found coincident descriptions and circumstantial details to support his new synchronization in so many ways made him confident of its correctness: "It would be a miracle . . . if all these coincidences were purely accidental. Anyone familiar with the theory of probabilities knows that with every additional coincidence the chances for another grow smaller . . . the chance would be a trillion or quadrillion against one that all the parallels offered on previous pages are merely coincidences. . . . However, we are not yet at the end of the journey"; it remained to carry the new chronology through to "the end of the last native dynasty in Egypt" [410:339–40]. This Velikovsky promised to do in volume 2 of *Ages in Chaos,* but publication of that work was continually delayed, and its (now several) volumes began to appear in print only in 1977 (see Chapter 5, *Velikovsky's Unpublished Works*).

The reception accorded *Ages in Chaos* was a muted echo of that given to *Worlds in Collision.* There was much less public fuss; some reviews were very favorable [105, 142, 283], others were scathing [4, 180, 197, 281]. Favorable reviews pointed to the massive documentation Velikovsky provided, with citations of the most varied sources, and to the "amazing, unheard-of, revolutionary, sensational" conclusions [142]. The negative comments derided Velikovsky's unorthodox synchronizations, in sarcastic tone [e.g. 180, 197, 281]:

> . . . Velikovsky has . . . an uncanny capacity for seducing the unwary. . . . [*Ages in Chaos* is] written with the air of a hopeful martyr. . . . Although he implies a knowledge of them, it is evident that he does not really know cuneiform . . . nor

> Egyptian, nor any other language necessary for the interpretation of the sources of the history of the ancient Near East, beyond the Hebrew and Aramaic which he learned in childhood. . . . [His method] is rather on a level with the possibly apocryphal story of the slightly addled New England professor who identified Moses with Middlebury by dropping "-oses" and adding "-iddlebury.". . .
>
> . . . a future historian might even more logically conclude that the Spanish Civil War of 1936 and the American Civil War are identical . . . an Abraham Lincoln Brigade fought in the former . . . [there is] no important person named Abraham Lincoln in Spanish history . . . the President of the United States at the time of the American Civil War was Abraham Lincoln. Hence, it follows that these two wars must be the same because of the identity of the names in question. As a rule, Velikovsky's own hypotheses are even more hazardous. . . . [4]

Earth in Upheaval

In *Worlds in Collision* and *Ages in Chaos* Velikovsky relies primarily on interpretation of legends and literatures. In *Earth in Upheaval* he reports hard facts of "stones and bones" that point to the occurrence of natural cataclysms of great magnitude in the history of the earth: the remains of mammoths apparently frozen rapidly, while standing and eating; mountainous assemblages of skeletons of many species; enormous ("erratic") boulders found hundreds of miles from their points of origin; evidence of sudden extinction of species. Most of these facts are quite well known; many (but by no means all) of them remain unexplained. In essence, Velikovsky recapitulates the evidence that had led scientists, up to the nineteenth century, to view the history of the earth as a succession of catastrophic occurrences, and he suggests that this evidence is so overwhelming that the more recent geological and biological view—that most changes have taken place over very long periods of time—ought to be abandoned. Beyond that, however, Velikovsky places many of the suggested cataclysms in the very recent past, in particular at about 1500 B.C. and about the seventh–eighth centuries B.C.—when, according to him, earth had close encounters with comet-Venus and with Mars respectively.

Again, some found Velikovsky's evidence impressive and could see merit in at least some of his interpretations [40], but others were outraged:

> "Earth in Upheaval" proves only that Dr. Velikovsky is a tireless library searcher and that libraries can be tricky places. They now contain so many books that a stubborn man can make a show of documenting almost any idea he cares to imagine. [103]

> Dr. Velikovsky is really a very intrepid scholar. In "Worlds in Collision" he sought to realign certain celestial chronologies. In "Ages in Chaos" he assayed a similar chore for historical and cultural timetables. Now he tackles the very sequence and timing of geochronology and the processes of organic evolution. Little by little, time and space are getting into step according to the good doctor's pace and cadence! . . . I have never doubted that Velikovsky reads widely—if not always wisely and well. . . . [198]

> . . . as generously packed with nonsense as were its predecessors . . . quotes some data which we know to be true, some which we know to be dubious and some which we know to be false. He quotes interpretations . . . which are correct and interpretations . . . which are wrong—and impartially he gives equal weight to each. . . . [38]

Oedipus and Akhnaton

> . . . belongs to a category of quasi-scientific fiction which Velikovsky has made peculiarly his own . . . collects the Egyptological data bearing on the personal and family life of early fourteenth-century Pharaohs . . . compares this . . . with details of the Oedipus stories as narrated by the Greek tragic poets of the fifth century B.C. Missing or uncertain elements in the Egyptian record are restored from the Greek tragedians, not without a great deal of violence to the Egyptological data. The patterns emerging . . . are finally interpreted in the light of Freudian theory. . . . All historical analogy is directly opposed

> to the ingenious structure of hypothesis . . . presented in Velikovsky's *tour de force* . . . good reading, but scarcely history. [5]
>
> . . . offering an explanation . . . that somehow managed to elude everyone for centuries, and is flat contrary to all the findings and most of the evidence . . . his persuasive arguments may carry a good deal of conviction to the uninitiated . . . a curious sense of unreality, an absence of common *nous*. . . . It is much to be feared that his ingenious theory, like so many others of this ilk, is moonshine. . . . [402]
>
> . . . his indiscriminate use of both secondary and primary sources . . . incurs the risk of taking probabilities or even mere possibilities for established facts. . . . Utilizing the known facts and the published material and bridging the gaps with specious hypotheses, he presents a fascinating drama which reads like a historical novel. . . . The book makes good reading, although many of Velikovsky's arguments and hypotheses cannot be accepted without further proof, and the abuse of Freudian concepts as a key to the understanding of antiquity has to be rejected. [454]

The Case for Velikovsky

So the opinion of professional scientists and historians was overwhelmingly that Velikovsky was quite wrong. Why did these judgments not carry the day in this public controversy?

That question is addressed at various points later in this book. The answers carry significant lessons for the present era, in which so many issues of social and political importance involve controversies engaging scientific and other experts as well as the public, the press, and governments. At this point in the story of the Velikovsky affair, one significant answer emerges: the virulence of the attack on Velikovsky, the victimizing of such individuals as Atwater and Putnam, and the boycotting of Macmillan repelled and appalled many people, who then sympathized with Velikovsky and were loath to side with those who attacked him. Velikovsky and his followers made the nature of the attack on him into a theme that was, and continues to be, used in his support: "If there is not truth in the work, why is so much ammunition directed against it? Are not many fantastic and worthless theories printed year in and year out, only to be met with silence?" [429]. Velikovsky suggested that "here is a rule by which to know whether . . . a book is spurious: Never in the history of science has a spurious book aroused a storm of anger among members of scientific bodies. But there has been a storm every time a leaf in the book of knowledge has been turned . . ." [410:foreword].

Velikovsky and his supporters recounted instances from the past in which the conventional scientific wisdom had roundly rejected new ideas that later were to win approval. Thus "uniformitarianism"—the idea that the earth's surface has been modified only slowly—was condemned as thoroughly unscientific, or worse, by then-conventional "catastrophists" before the middle of the nineteenth century; now the conventional uniformitarians were rejecting the new catastrophism of Velikovsky [142]. Giordano Bruno was burned at the stake for cosmological heresies [429]. Archimedes jeered at his elder, Aristarchus, who maintained that the earth traveled around the sun [431]. Sigmund Freud, rejected by the orthodox psychologists, spoke of the "brilliant example of the aversion to learning anything new so characteristic of the scientists" [429]. Simon Newcomb, one of America's greatest mathematicians, proved that a manned flying machine was an impossibility—in the same year the Wright brothers flew [431]. "All fruitful ideas have been conceived in the minds of the nonconformists. . . . The truth of today was the heresy of yesterday" [431].

Further, Velikovsky's critics were accused of not examining his work carefully enough to reach an objective conclusion. He had been criticized on the basis of very brief summaries, in popular magazines, before *Worlds in Collision* had even been published [142, 143, 410:foreword, 429]. The denunciations of Velikovsky's theories were also said to lack specificity, to be couched in generalities about conflicts with existing physical laws [142, 251]. This assertion gained credence when some critics [112, 207, 290] of Velikovsky explained that it was not worth the time and effort to analyze and refute him detail by detail: "The difficulty . . . lies in the volume of material required . . . to list, first, the particular scientific laws violated by Velikovsky's hypothesis. . . . Each of these laws should be explained, the evidence for them . . . evaluated, and it should be pointed out just how it conflicts with the Velikovsky hypothesis. A fairly complete textbook of physics would be only part of the answer to Velikovsky, and it is therefore not surprising that the scientist does not find the undertaking worth while" [207].

There had, then, been violent and emotional attacks without careful reading and without detailed criticism, directed at an educated man of considerable accomplishment: "he has also published a historical novel . . . and some professional papers on mental health and disease. In one of them (1930) he anticipated the electro-encephalic changes in epileptics. . . . In 1948, under the name 'Observer,' he wrote a series of fifty columns on Middle Eastern politics for the New York *Post*" [142].

That the attacks on Velikovsky were made emotionally and in an "unscientific" manner constitutes a case against some of his critics, not a case for Velikovsky; it is not really relevant to the question of whether his views are plausible. Moreover, analogies with similar treatment of people who were later found to have been right do not begin to establish that Velikovsky himself might later be vindicated. Nevertheless, in such controversies it does seem that many people choose sides on the basis of such nonsubstantive or irrelevant points. After all, everyone can appreciate the situation of an individual who stands virtually alone in the face of furious opposition, and there is a common inclination to side with the underdog. By contrast, few people are in a position to decide for themselves the technical merits of such complicated arguments. In effect, then, the manner in which Velikovsky was attacked constituted an avenue through which he could gain support, at least among the ranks of nonscientists. Certainly he and his supporters have continually striven to make this case against Velikovsky's critics into a case for Velikovsky.

In the exchange of polemics Velikovsky showed himself to be a persuasive debater, at least a match for his opponents. In contrast to the generally intemperate tone of his critics, he maintained a cool and confident manner, occasionally quite incisively sharp:

> If I had not been psychoanalytically trained I would have had some harsh words to say to my critics. [35]

> Here we have a good case of collective scotoma (blind spot), if one prefers that term [to "collective amnesia"]. It is the inability to recognize and properly read a great number of testimonies brought together between the two covers of a book. From the psychology of the unconscious mind we know that the amnesia of traumatic experiences is accompanied by emotional outbursts at an attempt to unveil it. We have witnessed explosions of "highly unscientific fury." For if I am right, even if only to some degree, those who greeted the book with threats of boycott have taken a position which may easily become untenable. The reputation of scientists, at least of the entire branch of American astronomers, is thus jeopardized by a wager that *Worlds in Collision* is a calculated attempt to mislead the scientists and the public. [429]

And how could the astronomers be so certain, when so many things were left unexplained by their existing theories? When there existed no satisfactory explanations for the behavior of comets, for the roundness of the sun, for the influences on auroras and radio reception of strongly charged particles from the sun? How could astronomers be so sure that there never had been and never could be comets as large as Venus, simply because one had not recently been observed by scientists? [429].

Velikovsky went further, charging that as the years passed science was incorporating his ideas without acknowledgment and even while continuing to brand his views as unacceptable [410:foreword]. Thus Velikovsky's skepticism about the applicability of the law of gravitation, and his insistence that electromagnetic forces had to be taken into account in explaining the behavior of the bodies in the solar system, had been dismissed by his critics. Payne-Gaposchkin in 1950 had written two major articles attacking Velikovsky's ideas and had said "that the celestial bodies 'could not possibly possess electrostatic charges enough to produce any of the [observed] effects on motion within the solar system'; . . . [but] in the September 1953 issue of *Scientific American*, [she makes] this confession: 'Ten years ago in our hypotheses of cosmic evolution we were thinking in terms of gravitation and light pressure. . . . Tomorrow we may contemplate a galaxy that is essentially a gravitating, turbulent electromagnet'" [431]. Velikovsky's scenario of cosmic collisions had been pooh-poohed; yet within a few years astronomers were suggesting that the asteroids had been formed by disintegration of a planet, or upon collision between two planets, or from the remnants of a gigantic prehistoric comet. Whipple presented evidence that the asteroids had collided with a comet twice in historical times, about 2800 B.C. and about A.D. 400; Vsehsviatsky thought that many comets could be of recent origin [431]. Velikovsky had been rejected in part because such phenomena were dismissed as impossible.

So Velikovsky counterattacked, and challenged his critics to put his ideas to the test of objective experiment: ". . . I have suggested various tests to cover one or another aspect of my theory. . . . I ventured the guess that the atmosphere of Mars consists of argon and neon . . . that petroleum gases must be present on Venus . . . methods exist to check on some segments of the theory . . ." [429].

The Case against Velikovsky

Some of the detailed points raised by critics of Velikovsky have already been mentioned (Chapter 2, *Reviews*). None of the critics regarded Velikovsky's rejoinders as adequate, and he was further criticized for arguing rhetorically rather than answering the substantive points raised. For instance, he implied that well-known people gave him credence, when that was not in fact the case:

> . . . Velikovsky implies strongly that Albert Einstein was beginning to understand Velikovsky's views and that the two men were close to agreement: "The late Dr. Albert Einstein during the last eighteen months of his life (November, 1953–April, 1955), gave me much of his time and thought. . . . We started at opposite points; the area of disagreement . . . grew ever smaller, and though at his death . . . there remained clearly defined points of disagreement, his stand then demonstrated the evolution of his opinion in the space of eighteen months." . . . This carefully worded statement, upon close analysis, clearly says nothing definite or significant—but it creates an impression upon the casual reader. [38]

This example of implication appeared in a book review in *Scientific American* and was perhaps particularly interesting to readers of that magazine, who might have seen, less than a year previously, "An Interview with Einstein":

> The subject of controversies over scientific work led Einstein to take up the subject of unorthodox ideas. He mentioned a fairly recent and controversial book, of which he had found the nonscientific part—dealing with comparative mythology and folklore—interesting. "You know," he said to me, "it is not a bad book. No, it really isn't a bad book. The only trouble with it is, it is crazy." . . . He . . . went on to explain . . . this distinction. The author had thought he was basing some of his ideas upon modern science, but found the scientists did not agree with him at all. In order to defend his idea of what he conceived modern science to be, so as to maintain his theories, he had to turn around and attack the scientists. I replied that the historian often encountered this problem: Can a scientist's contemporaries tell whether he is a crank or a genius when the only evident fact is his unorthodoxy? A radical like Kepler, for example, challenged accepted ideas; it must have been difficult for his contemporaries to tell whether he was a genius or a crank. "There is no objective test," replied Einstein.
>
> Einstein was sorry that scientists . . . had protested . . . about the publication of such a book. . . . The author . . . might be "crazy" but not "bad." [53]

Martin Gardner [112, 113:32–35] labeled Velikovsky a crank who was determined to rationalize and defend the orthodox Jewish interpretation of Old Testament history. But perhaps the most damaging charge brought by Gardner was that Velikovsky's propositions were not at all original: they had been put forward in very similar form by William Whiston in the seventeenth century and by Ignatius Donnelly in the nineteenth. Both wrote of an encounter between the earth and a comet. Whiston claimed that before this encounter, the year had been 360 days long and the month 30 (similar to suggestions made by Velikovsky), and he wrote about magnetic forces acting to slow the earth's rotation. Both Whiston and Donnelly drew evidence from biblical reports and from legends and myths the world over; Donnelly remarked—as had Velikovsky—on the fall of stones from heaven, accompanying Joshua's miracle, as being what one would expect if a comet came close to the earth. Velikovsky's brief footnote references to Whiston and Donnelly did not reflect the degree of similarity between Velikovsky's own argument and those earlier ones, particularly in Donnelly's *Ragnarok.*

Arguments on Details

How was one to decide whether Velikovsky was right? He had put forward an intricately detailed and integrated picture of physical and historical events spanning many centuries: a massively documented case. Many prominent individuals were impressed. On the other hand, critics with impeccable credentials held the scenario to be impossible. Velikovsky and his supporters reminded us that critics with impeccable credentials had often been in error where revolutionary advances in knowledge were concerned.

In this situation, one could form an opinion in one of two ways: by deciding which experts to believe and which to reject, or by evaluating the evidence for oneself. The first has pitfalls—experts have been wrong in the past, the whole scientific establishment has been wrong at times. What criteria could one use to decide whether this happened to be one of those scientific revolutions? And, in any case, taking opinions at second hand is hardly satisfying or completely convincing. The second way seems impossible: how can one evaluate for oneself these highly technical minutiae in umpteen different disciplines? Perhaps, though, one could look closely at the exchange of polemics on those matters, and see whether one side or the other was evading issues and arguing illogically. That at least seems worth trying, but at this stage of the controversy no clear-cut answer would emerge.

One can find a few small matters on which Velikovsky did not give an adequate reply to criticism—for example, to Neugebauer's challenge concerning the meaning of a German passage dealing with the Chaldean ecliptic [269]. And one can find small details on which Velikovsky seems to have the better of the argument—for example, that the size of presently observed comets does not mean that there have not in the past been larger ones, actual cases being cited by Velikovsky [429]. The arguments on really crucial points, however, were inconclusive.

For instance, is it physically possible that Joshua's miracle was an actual event? Velikovsky [408:40–44] accepts that the sun did appear to slow down or stand still, and suggests either a disturbance of the earth's rotation or a tilting of the axis. Payne-Gaposchkin ridiculed the possibility of a disturbance of the rotation but did not address the plausibility of a tilting of the axis [287, 289]. Stewart [385] tried to show how silly the notion of disturbed rotation was: "The author perhaps does not fully appreciate what a sensitive indicator the oceans would be. Try it with a full dishpan in the back seat of your car. . . ." It is not clear how this was supposed to contradict Velikovsky's thesis, which is that enormous tidal waves in fact ensued. But even further, the suggested experiment did not give the predicted result: "Stewart makes the flippant statement: 'Try it with a full dishpan in the back seat of your car.' I did, and it appears that I can accelerate two miles per hour in one minute without any spilling whatsoever. . . . This would mean eight hours to as low as one and one-half hours for complete deceleration of the Earth. . . . I may be all wet—but the car stayed dry" [61].

So Velikovsky's suggestion of a slowed rotation had not been demonstrated impossible; the axis-tilting possibility was ignored; and Stewart—and, by association, other professional scientists—was wrong about as simple an experiment as a dish of water in a moving car.

Then there was the question of ancient astronomical observations. The dates of early eclipses of the sun were said to be consistent with the present state of the solar system, so that the encounters postulated by Velikovsky could not have occurred. Eclipses were mentioned from 2137 B.C. [393], 1062 B.C. [289], and 726 to 719 B.C. [269]. But Velikovsky pointed out that such dates were arrived at by retrograde calculations: eclipses mentioned in early records (chiefly those of Ptolemy in the second century A.D.) were checked against dates when eclipses would have occurred if planetary and lunar motions had remained unchanged. This could not test Velikovsky's contention that those motions *had* changed [429]. Stewart [385] rejoined by referring to three eclipses before 687 B.C., attested in Greek, Babylonian, and Chinese records. In reply, Velikovsky [430] showed that he was thoroughly familiar with the relevant literature: the century in which the Babylonian eclipse occurred was still a matter of debate, and one authority regarded the ancient descriptions as being not of an eclipse but of the earth's movement through a cloud of meteoritic dust. For the Chinese eclipse, the place of observation was not precisely known; the date was calculated "on the authority of the astronomer Y-hang who lived a generation later. . . . In his day, in 721 B.C. an expected eclipse did not take place. Y-hang informed the Emperor that 'the sky changed the order of the motions which cause eclipses.' He explained that already in earlier times . . . the sky changed the course of the planet Venus." As for Assyria, neither the place of observation nor the day of the month was given: "By retrograde calculation an eclipse should have occurred on July 15, 763 B.C. . . . an Assyrian chronology was built [on this basis] by reconstructing the lists of the magistrates. However, it required a change of 44 years in Biblical chronology."

Who won the argument on that question?

4

The Second Engagement

Who shall decide when doctors disagree?
—Alexander Pope

The first battle, just described, was concentrated largely in the years 1950 to 1952. From 1953, for almost a decade, there was little public mention of Velikovsky; the publication of *Earth in Upheaval* (1955) and *Oedipus and Akhnaton* (1960) produced brief flurries of reviews pro and con, letters, and editorial comments, but nothing approaching the publicity of the early 1950s. Then in 1962 there was a new public outburst and the issues were much more general. It became a question of the place of science in society, of how scientific activity is and should be carried out, of the responsibilities of scientists, of the relevance of studies in the social sciences—all this because a number of people became convinced that at least some of Velikovsky's ideas had been proved valid.

The fuse was lit when *Science* published a joint letter [22] from the physicist V. Bargmann (of Princeton University) and the astronomer Lloyd Motz (of Columbia University):

> In the light of recent discoveries . . . we think it proper and just to make the following statement.
>
> On 14 October 1953, Immanuel Velikovsky . . . [said] in a lecture . . .: ". . . Jupiter is cold, yet its gases are in motion. It appears probable . . . that it sends out radio noises as do the sun and the stars. I suggest that this be investigated."
>
> . . . on 5 April 1955 B. F. Burke and K. L. Franklin . . .

> announced the chance detection of strong radio signals . . . from Jupiter. They recorded the signals for several weeks before they correctly identified the source. . . . This . . . came as . . . a surprise because radio astronomers had never expected a body as cold as Jupiter to emit radio waves. . . .
>
> On 5 December 1956, . . . Velikovsky . . . suggested the existence of a terrestrial magnetosphere reaching the moon. . . . The magnetosphere was discovered in 1958 by Van Allen.
>
> In . . . *Worlds in Collision* (1950), Velikovsky stated that the surface of Venus must be very hot, even though in 1950 the temperature of the cloud surface of Venus was known to be −25°C . . . by 1961 it became known that the surface temperature of Venus is "almost 600 degrees [K] . . . a surprise . . . in a field in which the fewest surprises were expected. . . . We would have expected a temperature only slightly greater than that of the earth . . . the temperature is much higher than anyone would have predicted."
>
> Although we disagree with Velikovsky's theories, we feel impelled . . . to establish Velikovsky's priority of prediction of these . . . points and to urge, in view of these prognostications, that his other conclusions be objectively re-examined.

A new wave of public debate followed the appearance of this letter. Eric Larrabee returned to the fray in the pages of *Harper's* with an article [211] entitled "Scientists in Collision: Was Velikovsky Right?" He made clear that, in his view, Velikovsky had indeed now been proven to be right and that the injustices done to Velikovsky "require of scientists an act of agonizing reappraisal." Larrabee was attacked by Donald H. Menzel (director of the Harvard College Observatory) for his

> irrational attack upon science and scientific method. . . . As an apostle of the New Science, Larrabee carries on where Velikovsky left off, quoting selectively from various "authorities" to fashion an argument that may well convince the nonscientist that Velikovsky was correct, after all. . . . There is no master scientist who, like Aristotle, Nostradamus, or Velikovsky, has merely to furrow his brow and come up with all the answers . . . the methods advocated by Velikovsky and Larrabee represent a return to the dark ages. They no more

> represent science than the practice of voodoo represents medicine . . . readers have a right to expect responsible journalism from their leading magazines. [247]

Larrabee's response was brief [212] and re-emphasized the accuracy of Velikovsky's predictions.

Meanwhile, some social scientists had entered the arena. The September 1963 issue of the *American Behavioral Scientist* was devoted entirely to the Velikovsky affair. In a foreword the editor, Alfred de Grazia, recalled

> acrimonious debate and bellicose maneuvers over the validity of the new historical and cosmological concepts of Dr. Immanuel Velikovsky. . . . What has not been appreciated . . . is the high involvement of the social and behavioral sciences. The social sciences are the basis of Velikovsky's work. . . . Yet social scientists have been generally unaware of his work and almost totally disengaged. . . .
>
> Of equal importance to behavioral science is the controversy itself. The politics of science is one of the larger social problems of the twentieth century. . . . The central problems are clear: Who determines scientific truth? What is their warrant? How do they do so? . . . In the end, some judgment must be passed upon the behavior of science and, if adverse, some remedies must be proposed. [71]

The journal contained three major articles. The first was by Ralph Juergens, "a civil engineer and an associate editor on the staff of a McGraw-Hill technical publication . . . long . . . interested in the Velikovsky case"; Juergens discussed the events in the controversy. Then Livio Stecchini reviewed the history of ideas about the stability of the solar system, from the days of the Greek philosophers to modern times; "Dr. Stecchini is an editorial associate of the *ABS* [*American Behavioral Scientist*] and author of a number of articles on the history and methodology of science. His particular interest is the origin and development of quantitative science. He teaches history at Paterson State College and philosophy at Rutgers University." Finally, the process by which new theories are accepted or rejected was discussed in "The Scientific Reception System" by Alfred de Grazia, "editor and publisher of the *ABS* and professor of government at New York University."

The debate was carried on in the correspondence and editorial columns of the *ABS* for more than a year [9–11, 57, 63, 64, 68, 72, 225, 316, 363, 392, 463, 466]. There was also a sharp exchange [11, 72, 173, 236] with the *Bulletin of the Atomic Scientists,* which had carried an article deriding both Velikovsky and the *ABS* for lending him its support. In 1966 the articles from the *ABS* appeared in book form [78]; included were additional essays by Juergens, Stecchini, and Velikovsky, and two appendices, one of them the letter by Bargmann and Motz, the other an attempted demonstration that Payne-Gaposchkin had deliberately misrepresented Velikovsky in her critical reviews.

This second engagement in the war over Velikovsky's work was carried on in newspapers, magazines, and books [16, 93, 184:115–16, 213:71–77, 396:26–34], and continued for a decade; in the early 1970s there were new developments, to be recounted later. The battle began with arguments about Velikovsky's advance claims. To what extent had they now been confirmed? How many of them? Did such confirmation validate his ideas? That was one theme. Another was that the controversy was significant irrespective of whether Velikovsky was right or wrong, be it in detail or in his overall theses—the general question was raised of how scientific activity is, and how it ought to be, carried out. These discussions included quite sharp attacks on science and on scientists, particularly because of the way in which Velikovsky had been treated: more information about that had come to light. The polemics at times were very heated indeed, and a few observers attempted—without obvious success—to persuade the contestants to discuss the matter in a more dispassionate way.

The Advance Claims

The letter of Bargmann and Motz had credited Velikovsky with correct predictions of radio emissions from

Jupiter, of the existence of a terrestrial magnetosphere, and of the high temperature on the surface of Venus. Larrabee [211] cited further predictions that had been validated, as did Velikovsky himself [433] and others [107, 130, 263, 300, 307, 390]. The critics argued that the predictions had not in fact been verified, or were verified in a manner that made the facts different from Velikovsky's prior descriptions, or that Velikovsky's successes were merely lucky guesses.

Radio Noises from Jupiter

Velikovsky had said [431] that "in Jupiter and its moons we have a system not unlike the solar family. The planet is cold, yet its gases are in motion. It appears probable . . . that it sends out radio noises as do the sun and the stars. . . ." Later [433] he added that "these claims were not made casually or in a veiled form. Some of my arguments for Jupiter sending out radio-noises can be learned from my correspondence with A. Einstein. I could add that if the solar system as a whole is close to neutrality, and the planets possess charges of opposite sign to that of the sun, Jupiter must have the largest charge among the planets. Rotating quickly the charged planet creates an intense magnetosphere."

Menzel [247], one of Velikovsky's most outspoken critics, rejoined: ". . . what about . . . Jupiter as a source of radio noise? Clearly, Velikovsky's idea of an active Jupiter that had recently expelled Comet Venus suggested the idea to him. Since the idea is wrong, any seeming verification of Velikovsky's prediction is pure chance. It has no more validity than do his thousands of other erroneous suppositions and conclusions. Moreover, Venus is also a strong source of radio energy. Why didn't Velikovsky pick that one, especially since it is the intense radio emission that led to the discovery of a hot surface?" Much later Walter Sullivan, a science writer who had not participated in the controversy, wrote about it dispassionately: "While Jupiter does produce radio emissions, they have no relationship to emissions of the type generated by the sun and other stars, as proposed by Velikovsky . . . " [396:26–34].

The Earth's Magnetic Field

Velikovsky [431]:

> The electromagnetic nature of the universe, deduced in *Worlds in Collision* from a series of historical phenomena, is supported by another series of recent observations. . . .
>
> (It is generally thought that the magnetic field of the earth does not sensitively reach the moon. But there is a way to find out whether it does or not. The moon makes daily rocking movements—librations of latitude, some of which are explained by no theory. I suggest investigating whether these . . . are synchronized with the daily revolutions of the magnetic poles of the earth around its geographical poles.)

Larrabee [211] recalled that "in 1956 . . . Velikovsky submitted a memorandum to the American Committee for the International Geophysical Year, suggesting that the Earth's magnetic field might be stronger above the ionosphere and have effects as far as the Moon. The discovery of the magnetosphere was made in 1958 by James A. Van Allen, whose name the belts of radiation he found around the Earth now bear."

Menzel [247] replied that "Velikovsky forecast nothing resembling the Van Allen radiation belt . . . the earth's magnetic field . . . outside of the ionosphere . . . actually . . . is weaker . . . [and] suddenly breaks off at a distance of several earth diameters. Velikovsky said nothing about the entrapped electrons and ions, which distinguish the Van Allen belts from an ordinary 'magnetosphere.' . . . The great Norwegian scientist, Störmer, had developed a quantitative theory of the magnetosphere and its entrapped radiation early in this century. . . ."

Larrabee [212]: "The magnetic field . . . does not break off suddenly 'at a distance of several earth diameters.' The satellite Explorer X made measurements to beyond 42 earth radii (*Science,* July 12, 1963) and found a relatively stable geomagnetic field out to 22 radii; the field reappears at greater distances. Störmer's theory 'concerned the formation of the aurorae and I do not think that any scientist really anticipated that the earth was permanently inclosed in such a shell of particles' (Sir Bernard Lovell, *Nature,* September 8, 1962). . . ."

Walter Sullivan [396:26–34]: "The earth's magnetic field, far from becoming more intense above the atmosphere . . . terminates in a well-defined boundary, the 'magnetopause,' about one quarter of the way to the . . . moon except directly opposite the sun, where it streams out into a long tail." "It is hard to see how his [Velikovsky's] supporters can relate this to the Van Allen . . . belts" [395].

The Temperature of Venus

Menzel [247]:

> As to the "high temperature" of Venus, "hot" is only a relative term . . . liquid air is hot, relative to liquid helium; the sun's surface is cold, relative to the star Sirius . . . Velikovsky . . . refers to actual astronomical observations of the infrared radiation from Venus, which showed that the dark side . . . was just as hot as the sunlit side. The measured temperatures were comfortably warm, not 800°F. . . . Velikovsky . . . even suggested that petroleum might still be burning on the surface of Venus. But—to fulfill another one of his hypotheses—Venus had to be cool enough to support insect life. . . . He can't have it both ways!

Larrabee [212]:

> The accepted figure for Venus's cloud-surface temperature . . . was very low (−25°C) when Velikovsky wrote that at the planet's surface it must be high; all hydrocarbons turned to gases is a fair example of what he meant by "hot." There is no inconsistency here with his tentative suggestion that there might also have been vermin in its "trailing atmosphere." Some features of Venus's . . . spectrum have been attributed to . . . micro-organisms by the Russian astronomer Kozyrev; and the director of NASA's Office of Space Sciences, Dr. Homer E. Newell, made precisely the same suggestion—that in its cooler atmosphere "a low order of life form" could exist. . . .

Regarding the high temperature of the surface of Venus, and Velikovsky's prediction thereof, Juergens [172] wrote that "no convincing explanation has yet been advanced to square this . . . with orthodox cosmologies."

However, according to Walter Sullivan, "although Velikovsky apparently was unaware of it, a decade earlier Rupert Wildt . . . had proposed that the surface temperature of Venus must be above the boiling point of water"; and "from what is now known of the Venus atmosphere it would be most remarkable if its surface temperature were not oven hot, for the planet is covered with a massive blanket of carbon dioxide . . . [which] traps solar heat and keeps both the day and night sides extremely hot—even hotter than predicted, from such a 'greenhouse effect,' by Rupert Wildt in 1940" [396:26–34].

Electromagnetism and the Earth's Rotation

Velikovsky [433]: "My thesis that changes in the duration of the day had been caused in the past by electromagnetic interactions was rejected in 1950–51. . . . In February 1960, A. Danjon . . . reported . . . that following a strong solar flare the length of the day suddenly increased by 0.85 millisecond. Thereafter the day began to decrease by 3.7 microseconds every 24 hours. . . . He ascribed [this] . . . to an electromagnetic cause connected with the flare. . . ."

Menzel [247]: "The explanation for Danjon's observation is simple: a minute temporary expansion of the earth, caused by heat from the flare, accounts for the slowing. It has no possible relation to the Biblical account. There is an enormous difference between one one-thousandth of a second a day and making the sun stand completely still."

Larrabee [212]: "Dr. Menzel's view . . . is contradicted by Dr. Danjon himself: 'It is very likely electromagnetism alone which will furnish the explanation . . . because the thermal effects involve only a tiny fraction of the earth's mass' (*Comptes rendus des Sciences de l'Académie des Sciences,* Volume 250, Number 8, page 1402, February 22, 1960)."

Thus, on a number of specific points, Velikovsky and his supporters claimed that he had been vindicated by recent actual findings, while the critics conceded neither on the facts themselves nor on how well they fitted the predictions. Some of these arguments have continued to the present time (see, for example, references 99, 119, 201). As the above excerpts illustrate, neither side of the argument was clearly compelling, and that largely remains the case.

Not all of the points raised as validations of Velikovsky's ideas were addressed by his critics; was he perhaps indisputably right in those instances?

History of the Near and Middle East

Velikovsky: "Texts in the Minoan (Linear B) script were found years ago on Crete and in Mycenae and in several other places on the Greek mainland. I believe that when the Minoan writings . . . are deciphered they will be found to be Greek. I also claim that these texts are of a later date than generally believed . . ." [431].

"This lecture [excerpt above] was delivered on October 14, 1953. In November of the same year the first announcement of the decipherment . . . was made by Michael Ventris, an English architect. Contrary to what had been thought . . . it was found to be in the *Greek* language. This . . . startled the scholarly world, as the texts had been erroneously referred to a time before the twelfth century . . ." [431:278n4].

On another point of his system of revised dates, Velikovsky also found justification:

> [Using] the radiocarbon method of dating . . . Wood from under the foundation of the "Hittite" fortress of Alisar . . . turned out to be seven to eight hundred years younger than conventional chronology would allow, thus giving full support to my dating. . . . The age of pieces of wood from the . . . Old and Middle Kingdoms in Egypt also proved to be in harmony with my reconstruction. However, for the decisive period—that of the New Kingdom—no radiocarbon analysis has been made. . . . Soon you will be able to judge as right or wrong my unqualified statement that carbon analysis of . . . the furniture and sacred boats of Thutmose III or Tutankhamen, would yield dates five to seven hundred years younger than those assigned by adherents of the conventional chronology. . . . [431]

One of Velikovsky's supporters amplified this point after such measurements had actually been made:

> Velikovsky's efforts of more than a decade to induce radiocarbon laboratories . . . to test objects from the New Kingdom . . . have yielded their first fruits. The test results are compatible with Velikovsky's chronology and quite incompatible with the conventional timetable . . . wood from the tomb of Tutankhamen . . . [gave a date of] 1030 ± 50, B.C. (. . . [or] 1120 ± 52, B.C.) . . . accepted chronology . . . places Tutankhamen in the fourteenth century. Velikovsky places him in the ninth . . . it is possible that heartwood grown about 1030 (or 1120) B.C. was cut in the ninth century . . . it is not possible, however, that wood grown centuries after his death furnished objects for a fourteenth-century pharaoh. [173]

Many other such vindications have been cited: evidence of early human culture *had* been found in Siberia; the end of the last glacial period *was* much more recent than previously supposed; pre-Columbian civilizations in Mexico *were* a thousand years older than previously believed; archaeological digs in the Middle East *had* found that "the entire area was shaken and devastated . . . by natural disasters of a scope and severity with which modern experience has nothing to compare . . . and precisely as Velikovsky maintained, ended the Middle Kingdom in Egypt . . ." [211].

Significance of the Advance Claims

Even if Velikovsky had indeed been right in some of his predictions, his critics claimed this did not prove the correctness of his theories. Thus Poul Anderson [15], a writer of science fiction, responded sarcastically to Bargmann and Motz: "Fair's fair, of course, but why stop with Velikovsky? In . . . Gulliver's . . . voyage, Swift remarked that the Laputans had discovered that Mars had two moons . . . a century and half before Hall found them telescopically . . . we ought to try out some of the other remarkable experiments and devices of Laputa. . . ." Anderson said he could equally find in what "purported to be only slightly fictionalized accounts of the author's [Richard S. Shaver] . . . former existence in vanished Lemuria" predictions of "solar particle emission . . . the aging and mutagenic effects of ionizing radiation . . . recent findings as to the effect of direct electrical stimulation of various brain centers. In view of these prognostications, his other conclusions must be objectively re-examined—unless, that is, one simply feels, as I do, that while one bad apple spoils the rest, the accidental presence of one or two good apples does not redeem a spoiled barrelful."

Another correspondent dismissed "Velikovsky's suggestion . . . about a terrestrial magnetosphere" as "in the nature of an ad hoc guess" and claimed that it could have been deduced on proper grounds from two of the correspondent's own publications [336].

According to Isaac Asimov [16], "If anyone reads *Worlds in Collision* and thinks for one moment that there is something to it, he reveals himself to be a scientific illiterate. . . . This is not to say that some of Velikovsky's 'predictions' haven't proved to be so. . . . However, any set of nonsense syllables placed in random order will make words now and then, and if anyone wants to take credit for Velikovsky's lucky hits, they had better try to explain the hundreds of places where he shows himself not only wrong but nonsensical." Asimov referred to the dating of the Papyrus Ipuwer, to the angular velocity of rotation of a planet's satellites, and to the composition of comets, among others.

Velikovsky, of course, believed that the vindicated claims were support for his work in general—"I have insisted in my published works, in my lectures, and in my letters that these physical conditions [referred to in the Bargmann-Motz letter] are directly deducible from my theory" [433].

According to Juergens [172], "Seldom in the history of science have so many diverse anticipations—the natural fallout from a single central idea—been so quickly substantiated by independent investigation. . . . Prof. H. H. Hess . . . Chairman of the Space Science Board of the National Academy of Science . . . wrote to Velikovsky: 'Some of these predictions were said to be impossible when you made them; all of them were predicted long before proof that they were correct came to hand. Conversely, I do not know of any specific prediction you made that has since proven to be false.'"

The arguments over Velikovsky's advance claims bear a certain resemblance to the earlier ones over his use of references and the possibility or impossibility of the physical events that he envisaged. On virtually every specific point something could be found pro and something con, and the arguments either became highly technical or veered off into generalities. Neither critics nor supporters would admit to error, so that there was no way for the casual observer to decide whether one side was getting the better of the other. Furthermore, so many other issues were brought in that, at times, the contestants did not even seem to regard the matter of the advance claims as necessarily decisive, nor the merits of Velikovsky's case the most significant issue.

Right or Wrong Is Not the Issue

The *American Behavioral Scientist* [71] commented: "From our point of view, at least the basic issue raised by the Velikovsky case has little to do with the correctness or otherwise of his theories. What is in question is the entire reception system that science uses in dealing with innovation." A number of people agreed:

> What is really at issue are the mores governing the reception of new scientific ideas on the part of the established spokesmen for science . . . there is the problem all editors face in discriminating between the work of a crackpot and the work of a genius . . . they are hard to distinguish, especially on the more advanced levels. . . . [225]

> [The] question is not whether Velikovsky is right or wrong, but the adequacy of the "scientific reception system" in our Age of Enlightenment. . . . [57]

> Even if he [Velikovsky] were wrong, . . . the problem is still as great. [316]

> A cultural fact of the present moment is the case of Dr. Velikovsky. The merits of the scientific issue do not alter the deplorable treatment that his ideas received from the profession. The pre-judging, scorn, and closing of ranks against the heretic showed typical guild animus. . . . [23:78n]

> The crux of the matter is not the validity of Velikovsky's particular historical interpretations, but whether an entire body of scientific evidence can be rejected on dogmatic premises. [378]

Lloyd Motz, whose letter with Bargmann had stirred up this hornet's nest, wrote to *Harper's* to clarify his position, which had perhaps been misinterpreted:

> . . . a careless reading of Eric Larrabee's article may leave the unwary reader with the false impression that Dr. Bargmann and I accept and agree with Dr. Velikovsky's ideas . . . I do not support Velikovsky's theory but I do support his right to present his ideas and to have these . . . considered by responsible scholars and scientists as the creation of a serious and dedicated investigator and not the concoctions of a charlatan seeking notoriety.
>
> . . . Dr. Velikovsky's ideas do not constitute a new theory since they contain no new fundamental principles of nature. . . .
>
> That there is no astronomical evidence for electromagnetic forces of the magnitude required by Velikovsky's theory . . . and that such forces of the required magnitude . . . would destroy the . . . completely verified laws of planetary motion are not accepted by Dr. Velikovsky as valid arguments against his ideas. Since . . . these . . . have led him to certain predictions . . . he is convinced that his ideas must be right. But . . . verified predictions alone do not validate a theory, and my position is that nothing has happened during the last decade to make Velikovsky's theory any more acceptable now than . . . when . . . first published . . . .
>
> . . . however, . . . his predictions should be recognized and . . . his writings . . . carefully studied and analyzed because they are the product of an extraordinary and brilliant mind, and are based on some of the most concentrated and penetrating scholarship and research of our period. . . . Dr. Velikovsky

> has performed a service to science in collecting the vast amount of data . . . and bringing clearly to the attention of the scientific community the many discrepancies that exist in our understanding of the history of our earth during the last geologic period. [260]

It was being said, then, that irrespective of any particular merit in Velikovsky's case, the controversy pointed to an inadequacy of the procedures by which new ideas are examined by scientists. Motz claimed that respect was due Velikovsky for his scholarship even though his approach and conclusions might be scientifically unacceptable. Horace Kallen [209] had expressed a similar view, as did David Stove [389]: "Most people engaged in intellectual work know how extremely few men are lucky enough ever to get a *new* idea, however small. The man who gets a new idea, and a huge one, rich in implications and new perspectives, and not known to be absolutely impossible, is one of nature's rarest prodigies. Such is Velikovsky. Test his originality this way: suppose . . . [his theory] is true—*then with whom would you compare him?* And then remember that truth is not in question when we are discussing originality."

So some people believed that larger issues were at stake than the specific one of whether or not Velikovsky was right. Nevertheless, as one reads these discussions, one finds that these same people, with few exceptions, did believe that Velikovsky was right, to some extent at least. For example, *ABS* had stated [71] that the basic issue was not whether Velikovsky was right; at the same time, the editors clearly believed that he was: "The probings of spacecraft tended to confirm—never to disprove—his arguments . . ." [78:2]. "Thirteen years ago astronomers were unanimous in dismissing as preposterous Velikovsky's contention that the movement of the heavenly bodies is affected by electromagnetic fields. Today creative astronomers are immersed in the study of electromagnetism. . . . Velikovsky saw what other scholars were not able to see . . ." [378]. ". . . Velikovsky is a fine scholar . . . a great cosmogonist . . . many scientists . . . would do well . . . to see what they may learn from his books" [11].

Alfred de Grazia, writing at length about the reception of new ideas in science as exemplified by the affair, surely thought that much was right in Velikovsky's work:

> . . . The works of Velikovsky are actually high in the scale of adduced proof and formality, by the standards of all past useful scientific production. . . .
>
> . . . his findings appear to be increasingly validated. . . .
>
> . . . The scope and importance of the knowledge involved are great. . . .
>
> Immanuel Velikovsky propounded a synthetic theory of the highest order. . . . He . . . has given us a new understanding of man's nature. . . . With rare imagination and consummate skill, he fashioned . . . theories of great scope, compactness, and integration. While his ideas are not at all beyond criticism, as a cosmogonist he appears in the company of Plato, Aquinas, Bruno, Descartes, Newton and Kant. What would therefore be only the duty of the critics of science—to defend ordinary or even mistaken scholars—becomes, by accident, an occasion to defend a great savant of the age. [73]

The attempt to leave judgment of correctness aside ("his ideas are not at all beyond criticism"), in the face of a presumptive belief that Velikovsky was right in many ways, led to some apparent contradictions. Compare "By the use of the methodology of social science . . . Velikovsky launched his challenge. . . . No one pretends that this method is adequate" [78:2] with "a more broadly educated or at least philosophically trained scientific class would have been able to perceive the relevance, validity, and unique capabilities of Velikovsky's method to key problems of natural science . . ."; and "Velikovsky's method . . . is . . . not highly self-conscious, *not always* exposed. It is much more clearly recognizable to social scientists than to natural scientists. Sometimes the method is concealed by an easy style that separates empirically-tied ideas, while allocating them to short sentences. Of course, a number of the rational propositions, which lend the work its distinction, are only as explainable as the leaps of Poincaré and Gauss . . ." [73].

A number of social scientists and humanists believed that Velikovsky had been vindicated: "[In] our own time Immanuel Velikovsky . . . appears to be approaching vindication . . ." [138]. "The image shaped of his findings . . . is a reasoned picture of events in the solar system recent enough for men to have observed and recorded in historic times . . . a signal achievement of imagination and intelligence working in close harmony to restore authenticity to an early image of an actual happening . . ." [184:115–16]. Doubleday put it this way [93]: "With the discoveries of the International Geophysical Year and the dawn of the space age, a number of his major predictions have been confirmed. . . . To many . . . the old question—'Could Velikovsky be right?'—became: 'How could such a heretic have known?'" Other support came from possibly unwelcome sources: "Science now tends to confirm the ideas on earth changes in the Edgar Cayce readings and in the writings of Dr. Immanuel Velikovsky . . ." [234].

It is worth noting that those who now believed Velikovsky to have been proved right had been impressed by, and quoted as evidence, the advance claims that had purportedly been fulfilled. And yet Motz and Bargmann, the scientists who had catalyzed this new phase of the controversy by suggesting that Velikovsky be given credit for his predictions, still believed Velikovsky's ideas to be basically incorrect.

Right about What?

While most of the public contestants talked about Velikovsky being "right" or "wrong," a few did mention the possibility of something between these extremes. The philosopher David Stove [389] pointed to the lack of success met with by scientists who had tried to prove conclusively that Velikovsky was wrong, and suggested that part of their problem was a lack of clarity as to precisely *what* they were trying to prove wrong—just as Velikovsky and his supporters had not always made clear precisely what elements of his views they claimed as now substantiated:

> In order to avoid supposing that astronomers have tried to persuade the public that something is impossible which they

know is not, we must . . . distinguish within Velikovsky's theory a number of theses, as follows:

(I) A thesis of general catastrophism: there have been sudden major changes in the physical state of the earth due to agents not observed to operate at present.

(II) A thesis of *extra-terrestrial* catastrophism; i.e. thesis I plus the clause that some of these agents have been extra-terrestrial.

(III) A thesis of *historical* extra-terrestrial catastrophism; i.e., thesis II plus the clause that some of these extra-terrestrial catastrophes have taken place in historical times.

(IV) The thesis of *Worlds in Collision:* i.e. thesis III plus the clause that one of these catastrophes was mainly due to comet-Venus, around 1500 B.C. . . .

Stove then reviewed the evidence in Velikovsky's books, as well as the advance claims, and concluded that "for theses III and IV . . . Velikovsky's harvest of evidence is meagre," though somewhat stronger for III than IV; there was worthwhile evidence to support thesis II.

On the face of it, of course, the battle was being waged over the correctness of IV, but Stove's analysis would help to explain some features of the controversy: "I think . . . we . . . have . . . the key to the violence of the astronomers' reaction. If . . . theses II and III open up immense possibilities, while there is nothing known to exclude them, then the uneventfulness of the history of the solar system is an assumption on which astronomers have placed a tacit reliance it by no means ever deserved. In the house that they knew so well, they had never noticed this door. And Velikovsky did the most infuriating thing in the world: he—a stranger—walked through this open door."

Something Amiss in Science

During this second engagement of the war the misdeeds of scientists were rehearsed in colorful detail. Some scientists seemed determined to provide evidence of their own misconduct. Two astronomers, Herget and McLaughlin, acknowledged with apparent pride [172] their parts in the boycotting of Macmillan books, McLaughlin emphasizing that he had not read *Worlds in Collision* and did not intend to. Doubleday, having taken over publication from Macmillan, had also been besieged: Whipple, Shapley's successor at Harvard, withdrew a popular series of astronomy books from one of Doubleday's subsidiaries [73]. The American Association for the Advancement of Science discussed ways of censoring books dealing with science, in order to prevent publication of such misleading works [378]. Shapley persuaded *Newsweek* not to publish an article on the apparent verification of Velikovsky's advance claims [173]. *Worlds in Collision* was suppressed in Germany too [172]. Evidence was adduced that Payne-Gaposchkin had deliberately misquoted Velikovsky in order to discredit him [73]. *Sky and Telescope,* a Harvard journal for amateur astronomers, deleted four pieces of data when publishing a report on the findings from the space-probe Mariner II to Venus—items that could be seen as lending plausibility to Velikovsky's claims [173].

There is more: officers of the Alfred P. Sloan Foundation campaigned against Velikovsky [225]. I. Bernard Cohen, historian of science at Harvard, was persuaded to revise remarks that could be construed as favorable to Velikovsky [73, 172, 233]. The American Philosophical Society refused to publish a paper by Velikovsky even though it was recommended by a prominent member of the society, H. H. Hess, who was also president of the American Geological Society [173]. *Science* and *Scientific American* refused to accept publisher's advertisements for Velikovsky's books but would not put their refusal in writing [173]. *Scientific American* would not publish Velikovsky's rebuttal of an unfavorable review.

Science had published the Bargmann-Motz letter that set off the second engagement, as well as a sarcastic reply to that letter (Chapter 4), but would not print Velikovsky's response. In explaining that decision, the editor of *Science* wrote:

> Science can exist and is useful because much of the knowledge in it is more than 99.9 percent certain and reproducible. If science were based on suggestions that were true 50 percent of the time, and all were free to make predictions which were only that reliable, chaos would result. I have repeatedly seen

> men of brilliance with fertile imaginations make all kinds of suggestions. Ideas are easy. They are cheap. It is the proving of a suggestion beyond a reasonable doubt that makes it valuable. . . . At least half of Velikovsky's ideas have been proved wrong and he has done little to substantiate the remainder. In view of this, he is not to be taken seriously. [73]

The astronomer Donald Menzel had calculated in 1952 that Velikovsky's scenario required the sun to have an electrical potential of 10^{19} volts, stated to be an impossibly high value. But in 1960 V. A. Bailey, emeritus professor of physics at the University of Sydney, proposed a new theory that had as one consequence an expected solar potential of 10^{19} volts (Bailey had been entirely unaware of Velikovsky's work and Menzel's calculation). In 1963 Menzel, by then director of the Harvard Observatory, responded intemperately to the suggestion that Velikovsky's advance claims had been validated, and he asked Bailey to retract his hypothesis, which was giving aid and comfort to Velikovsky's supporters. Bailey, however, found an arithmetic error in Menzel's calculations, which did not prevent Menzel from continuing to attack Velikovsky, partly on the basis of the incorrect calculation [173].

The Franklin Institute in Philadelphia had, since 1929, provided meeting rooms for the Rittenhouse Astronomical Society. When that society in 1967 invited Velikovsky to give a lecture, the institute refused to permit its room to be used. Much was made of this in the local press, and Velikovsky's talk was given in the Free Library, across the street from the institute. The Rittenhouse Society had about 350 members, but Velikovsky drew an audience of some 450 [233, 381].

So there is ample evidence that a number of scientists sought to prevent Velikovsky from publishing books and articles and even from giving lectures. A major theme in the whole controversy was: to what extent might this treatment of Velikovsky be symptomatic of something generally "rotten in the state" of science? —a situation of importance and significance beyond the particular case of Velikovsky. Was it still possible nowadays that scientific hypotheses could be criticized and suppressed on dogmatic grounds—as heretical in relation to the conventional wisdom of science—so that revolutionary, possibly worthwhile ideas would not be granted a fair, objective hearing? A number of people continue to believe that the Velikovsky affair shows that, indeed, it could happen, here and now.

The most comprehensive indictment of science came from de Grazia [73]. Scientists, he said, pay lip-service to the "rationalistic" model of science, in which there is freedom to publish new ideas and results that are then discussed and accepted or rejected on their merits. The Velikovsky affair demonstrated, however, that scientists behave in practice as though they dogmatically assert the truth of established knowledge, held on faith. They sought to prevent free publication, they attacked Velikovsky without having read his work, they refused to make tests of his theories.

Attacks were launched on the whole "establishment" of science: it "operates much like other social institutions, complete with hierarchy, dogma, and coercive power. Truth tends to be confused with orthodoxy" [46]. "We are back at scholasticism . . ." [378]. "I was deeply shocked, though not surprised, when the Scientific Establishment ganged up on Dr. Velikovsky. . . . the behavior of Messrs. Shapley, Mather, Condon and Haldane struck me as outrages . . . denying to others those very academic liberties which they professed to cherish" [392].

De Grazia and his co-editors called for action "to reform the errors of the vast enterprise of science" [78:4], and made specific suggestions: "radicalism in method" should not be "a deterrent to the recruitment of ideas" [73]. Natural scientists should systematically study "what the records of antiquity can contribute to the natural sciences" [378]. ". . . the Velikovsky record" should be read by those considering revisions of curriculum for students of science, and that curriculum "must be broadened to include a knowledge of the aims and methods of the humanistic and behavioral disciplines" [73].

An institute was said to be needed for research on the

behavior of scientists. Psychiatric techniques should be used "to give specialists insight into their motives and behaviors." Work in all scientific disciplines should be surveyed and assessed, and imbalances corrected. Scientific societies and journals should be reformed; professional reviewers' associations should be set up. Procedures were needed "for trying and sanctioning unprofessional practices amongst professionals," practices impinging both on ethics and on nonrationality:

> Perhaps Harvard University has within its authority the right to inquire into the scientific behavior of its faculty. Its officers might make a determination "on the merits" that one or more members of the faculty were so irrelevant and destructive in their scientific work as to violate plain standards of scientific competence. They might . . . take remedial action . . . to require apologies, . . . discussion in open forums, suspension, reprimand, resignation, or dismissal. . . . Scientific associations might conduct the same kind of inquiries. . . . Research is needed . . . into the conditions under which a hearing procedure . . . can be structured independently of the organization as a whole, very much as an independent court system operates in civil law. . . . The question arises . . . whether the larger society should ever take a hand in professional affairs. . . . Legislative and executive machinery should be avoided as far as possible, but quasi-judicial machinery encouraged. Scientists have on the whole tender sensitivities. A mild exposure and embarrassment usually have great corrective value for them. [73]

A common complaint was that scientists are narrow specialists who reject interdisciplinary work per se, yet such integrative, interdisciplinary work is badly needed [73, 78:1–4, 389, 463, 466].

Summing Up the Second Engagement

The reader is in a position to judge from the given quotations the tone of the criticisms made of Velikovsky, and of those who criticized the critics. Intemperate language, *ex cathedra* statements, unsupported generalizations, argumentation *ad hominem,* evasiveness, and sophistry are to be found aplenty on both sides. A few individuals [57, 363] urged that such emotional outbursts be eschewed, but their pleas fell on deaf ears [9]. As one result, arguments were often sidetracked into irrelevancies. For example, Velikovsky had drawn an inference from something in Herodotus in a way that could be seen as quoting directly rather than as drawing an inference [408:81]. He was attacked by Payne-Gaposchkin [290] and Sprague de Camp [67] for misquoting, and the argument was carried on not over the substantive question of what could be inferred from the sources but over who had been more misleading, Velikovsky or his critics [10, 68, 78:251–55]. A particularly vitriolic and unproductive exchange occurred between *ABS* and the *Bulletin of the Atomic Scientists;* there were arguments over punctuation marks omitted in quotation, mis-spellings were castigated as though some major error had been committed, lawsuits were threatened [11, 72, 173, 236].

Velikovsky's supporters and critics were unable or unwilling to carry on a dispassionate dialogue directed toward the substantive issues. The failure of either side to concede reasonable points to the other continued to make it difficult, if not impossible, for the layman to reach an informed judgment on the merits.

Velikovsky seemed to have been proven right on some major points of ancient chronology—the age of Mesoamerican civilizations and the dating of Tutankhamen in particular. Could he be right about those things and yet wrong about so many others that were based on the same premises? Quite possibly, it seems to me. Even if his use of memories and records of catastrophes turns out to be a valid method of synchronizing various events, that alone provides no evidence that the catastrophes were caused by cosmic agents, nor for the particular assertion that those agents were Venus and Mars. It was even less clear whether Velikovsky's advance claims about physical conditions on Jupiter and Venus and throughout the solar system (electromagnetic fields) had been vindicated or not. And even if they were, these conditions could all be given alternative explanations in terms of orthodox theories.

One is left with the nagging question whether someone who is *basically* incorrect could so frequently be right in specifics (or even so nearly right) in so many disparate fields.

5

The End of the Beginning

> This is not the end. It is not even the beginning of the end. But it is, perhaps, the end of the beginning.
>
> —Winston Churchill

A New Climate of Opinion?

> Velikovsky's address before a large audience of scientists and engineers at Harvard (Feb. 17, 1972) was . . . [reported]. It was at Harvard, of course, that Velikovsky had faced his most intense and bitter opposition, and this lowering of the barriers . . . signalled the onset of what may well be the last phase of Velikovsky's battle with the "scientific reception system." . . . *Science* . . . and the *Bulletin of the Atomic Scientists* reversed long-standing policies by accepting . . . advertisements for "Immanuel Velikovsky Reconsidered." Further, the respectable British journal, *New Scientist,* published a paper urging the community of scientists to take a hard look at Velikovsky's work. [297]

The *Bulletin of the Atomic Scientists* had published in 1964 an article that ridiculed Velikovsky and his supporters [236]. In 1975 a professor of physics published in that journal an article entitled "Velikovsky: Paradigms in Collision" [195], in which the possibility was left open that Velikovsky might be correct.

This was apparently a very different atmosphere from the virtually monolithic opposition to Velikovsky in academe in the 1950s and the heated battle between critics and supporters during the 1960s. By the mid-1970s Velikovsky had given talks at the Ames and Langley Research Centers of NASA and at many colleges and universities—Brown, Carnegie Institute, Dartmouth, Duke, Furman, Hunter, Oberlin, Rice, Temple, Yale; the universities of New York (Buffalo), North Carolina, Pennsylvania; Washington University (St. Louis). Courses dealing at least in part with Velikovsky's ideas were being taught in many places, including Antioch, Franklin and Marshall, Oberlin; at the universities of California (Berkeley), Florida, New York (Buffalo), Tennessee, Toronto, Victoria; at Glassboro State College, California State University (Long Beach), Moore College of Art, Oregon College of Education; at Indiana, Michigan State, New York, North Texas State, Oregon State, Princeton, Seton Hall, Texas Christian universities. Velikovsky had been awarded an honorary degree by the University of Lethbridge in 1974. Television documentaries about him had been shown in Canada, England, and Holland.

In 1950 a planetarium feature based on Velikovsky's scenario had been canceled before its first showing. In 1970 the Reading School District Planetarium (Pennsylvania) ran such a program, entitled "Whence Cometh Venus," for nearly a month. In 1972 the Kendall Planetarium (Oregon) had a Velikovsky program, as did the Nobel Planetarium in Fort Worth, Texas.

Symposia devoted to Velikovsky's work and its implications had been held at Lewis and Clark College, Oregon, in 1972; at Lethbridge and McMaster universities in 1974; under the auspices of the Duquesne History Forum (Pittsburgh), the Philosophy of Science Association (Notre Dame), and the American Association for the Advancement of Science (AAAS) at its meeting in San Francisco; and at the Saidye Bronfman Centre in Montreal in 1975 [314]. Ten issues of a magazine were devoted to the Velikovsky controversy (*Pensée* [291]), and new journals of Velikovskian studies were founded (*Kronos* [199], *Review of the Society for Interdisciplinary Studies* [346]).

All this could be taken to indicate that in the period of about a decade, beginning roughly in 1962, Velikovsky had moved from being ignored, condemned, and *persona non grata* in academia to being accepted within the mainstream of academic life and discussion. However, such a conclusion, explicitly stated by a few of Velikovsky's supporters, is oversimplified and therefore misleading.

Up to the early 1960s Velikovsky had stood essentially alone, supported on occasion—regarding one matter or another—by a few individuals. Beginning with the intervention of the *American Behavioral Scientist* in 1963, there came into existence a more coherent group of supporters. Few in number, they nevertheless were able to decrease Velikovsky's isolation from the academic community by arranging lectures, symposia, and avenues for publication of discussions of his ideas. At the same time, fewer natural scientists chose to comment publicly and vitriolically on Velikovsky's ideas. It seemed to be generally accepted that Velikovsky had been treated badly even if he was quite wrong. Thus the American Association for the Advancement of Science arranged a symposium to provide a forum for dispassionate discussion of Velikovsky's work: the very association whose journal, *Science,* had in earlier years declined to publish articles by Velikovsky. Further, a few scientists and other academics in good standing were prepared to expend time and effort in exploring, in relatively objective fashion, the substance of some of Velikovsky's contentions.

All this, however, does not amount to acceptance of Velikovskian ideas into the mainstream of astronomy, history, evolutionary biology, or any of the other recognized disciplines. The latter proceed as before, just as though Velikovsky's work had never appeared. But side by side with the mainstream there runs a Velikovskian stream, with its own journals, pursuing its own direction. There is little mingling. Velikovsky is rarely mentioned in the older, established journals of science; when the mention is favorable, it at most asks that his work be examined in a spirit of fair play. Those articles that discuss Velikovsky's work in detail, and that are based on the premise that there is at least some truth in his work, are read at specifically Velikovskian symposia and are printed in Velikovskian journals.

Velikovskian Science and Its Practitioners

Velikovskian science is based on the assumption that Velikovsky's cosmic scenario is largely correct: Venus was born out of Jupiter, passed close to earth with catastrophic effect several times around 1500 B.C., sent Mars into close encounters with earth in the eighth and seventh centuries B.C., and is now a planet whose characteristics reflect that violent history. The time scale of historical and prehistorical events has to be changed from the conventionally accepted one, particularly for the Mediterranean cultures. The catastrophes caused by Venus were only the latest of a number of similar cosmic events. Their effects were important to geological and biological evolution, not least to the psychological development of mankind. We all have powerful, albeit unconscious, memories of those events—we suffer from "collective amnesia," the trauma of those catastrophes being so great that mankind suppressed explicit recollections and now remembers only in allusive and elliptical ways through myth, legend, and folklore. Careful analysis of the latter, however, together with evidence from, for example, archaeology and paleontology, makes it possible to reconstruct those catastrophic events.

This view has implications for literally every field of human learning, and Velikovskian science covers that whole range. Some Velikovskians work in rather narrow areas, fleshing in details of points made by Velikovsky. Others work to reconcile conventional disciplines—physics, history, biology, archaeology in particular—with Velikovsky's scenario. Still others apply Velikovsky's method to the analysis of other legends to elucidate characteristics of earlier catastrophes. And some Velikovskians write globally about the implications of the new science for our world-view [74, 240, 264, 468]. Increasingly, "revisionist" work is appearing: consonant with most of Velikovsky's method and general approach but differing with his conclusions, sometimes in major ways.

So papers at Velikovskian symposia and in Velikovskian journals have dealt with the most diverse matters: planetary orbits [334, 342–45, 368, 369]; whether earth was once without a moon [286, 423]; the magnetism of lunar rocks [403]; when the latter were last molten [418, 473]; Venus's atmosphere [44, 319, 424, 425]; whether Venus's temperature is decreasing [420]; radiometric methods of dating past events [220, 263, 377]; the bearing of results from those methods on Velikovsky's scenario [14, 45, 156, 220, 231, 263, 301, 377, 438, 471]; electrical and magnetic effects [41, 58, 62, 107, 175–78, 203]; the origin and chemistry of manna [205] and of coal [294]; Hawkins's [144, 145] interpretation of the alignments at Stonehenge [232, 419]; the archaeology and history of the Middle East [43, 156, 157, 265, 295, 298, 386, 387, 422, 437, 442–44, 446, 447] and of Mesoamerica [266]; Atlantis [131]; myth and the origin of religion [81]; the origin of the Chinese dragon as an actual picture in the heavens [397]: "a writhing, bright, elongated thing . . . irregular in outline; . . . apparently on fire. . . . This thing, the dragon, seemed to be driving off the terrible flaming globe and so to be benevolent as well as powerful."

Pensée [291] was published from 1972 to 1975, with a regular circulation of between 10,000 [117] and 20,000 [31], but the first issue in the Velikovsky series was reprinted twice; *in toto* some 75,000 copies were printed. In 1974 there appeared briefly a new publication: *Chiron—The Velikovsky Newsletter,* based at the Oregon College of Education; apparently only one issue was published [330]. An English group interested in catastrophism in general and Velikovsky in particular is the Society for Interdisciplinary Studies [346]; the membership is not large, but its *S.I.S. Review* gained wider circulation through the services of the Research Communications Network [337] and mentions in *Kronos* [199]. The latter, a quarterly, is now the chief Velikovskian organ; about a year after its inception in 1975, *Kronos* had a subscription list of about 1,000, from some 10 countries [350], and the list has now grown to about 1,500.

Contributors to these journals range from orthodox specialists who expound some aspect of their specialty, at times in criticism of Velikovskian ideas, through established specialists who keep an open mind about various aspects of Velikovsky's scenario, to the real aficionados. There are people with impeccable credentials here; see, for example, the staff of *Kronos,* which was listed together with professional identifications in each issue of the first five volumes of that journal: anthropologists, philosophers, physicists, psychologists, and others holding responsible positions in and outside academia.

Apparently the first organized Velikovskian group began with the Cosmos and Chronos study and discussion seminars at Princeton in 1965 [421]. There seems to be little such activity on campuses nowadays, but Cosmos and Chronos continues as "a tax exempt non-profit corporation . . . to help with the investigation and dissemination of information about Velikovsky's theories . . ." [331:cover]. The executive director, C. J. Ransom [330], offers illustrated lectures about Velikovsky to interested groups; his book, *The Age of Velikovsky,* was published in 1976 by Kronos Press.

Altogether, it appears that not a large number of people are engaged in Velikovskian research and discussion, but those who are seem to be quite active. In 1975 a Center for Velikovskian and Interdisciplinary Studies was established, affiliated with the Department of History at Glassboro State College.

Scholarly Support

While the journals of science remain silent about Velikovsky, a few respectably established scientists have attempted unbiased evaluations of what implications accepted physical laws might have for the plausibility of Velikovsky's postulated events. In 1950, and for years thereafter, many astronomers in particular had called those events impossible: contrary to the laws of Newton and of energy conservation, among others. Later, however, Velikovskians derived comfort from statements by professional scientists that those events are by no means impossible.

Irving Michelson, of the Department of Mechanics and Mechanical Aerospace Engineering at the Illinois Institute of Technology, discussed classical mechanics and possible electromagnetic effects [117, 248–49], concluding that Velikovsky's contentions "are certainly not at variance with classical mechanics." Further, the energy required to reverse the earth's axis (in relation to the stars) "happens to correspond closely to modern estimates of the energy of a single moderately strong geomagnetic storm. . . . Is it possible that Herodotus' baffling allusions to the 'Earth turned upside down,' as reported in Papyrus Ipuwer also, was triggered by a geomagnetic storm . . . ?"

One of the earliest criticisms of Velikovsky's scenario had been that "an approximate arithmetical regularity called Bode's law exists with respect to the distances of the planets from the Sun. . . . The orbit of Venus is just the size to be expected under this rule" [385]. Consequently, if Venus had not occupied its present orbit until the seventh century B.C., Bode's law would not have been true at earlier times—presumably a very implausible supposition to make. Velikovsky had replied that "Bode's law is but an observation of the arithmetical relationship between the distances of the planets from the Sun. . . . With Neptune and Pluto 'the law breaks completely.' No physical reason or dynamical principle was ever afforded for Bode's law. Gravitational theory cannot explain it" [430].

In 1972 M. M. Nieto discussed the law in question in a book [275] and also wrote a short summarizing article for *Pensée* [276]. The original law was of the form

$$R_n = 4 + 3 \times 2^n$$

with $n = -\infty, 0, 1, 2 \ldots$ ($-\infty$ for Mercury, 0 for Venus, 1 for earth, and so on) and R_n the distance of the respective planet from the sun. Nieto showed that a better fit for the observations—including Neptune and Pluto—was given by

$$R_n = 3.34\,(1.73)^n$$

with $n = 0, 1, 2, 3 \ldots$.

More or less concurrently, Ransom [309] showed that Bode's law did *not* require Venus to be in its present orbit: if the original law was rewritten as

$$R_n = 4 + 6 \times 2^n$$

with $n = -\infty, 0, 1, 2 \ldots$ ($-\infty$ for Mercury, 0 for earth, 1 for Mars, and so on), then it described the present state of the solar system but with Venus absent.

Bass [24, 25] found that Bode's law could be derived on the principle that the planets take up distances from the sun so as to minimize the total potential energy of the system: thus there *is* a physical and theoretical basis for the law. But, for this very same reason, if some massive new object were to enter the solar system and become a planet, then all the planetary orbits would readjust so as to maintain a Bode-type relationship. The existence of Bode's law was therefore an argument for the possibility of Velikovsky's scenario, not against it. Bass concluded that "it is perfectly possible, according to Newton's Laws of Dynamics and Gravitation . . . for planets to nearly collide and then relax into an apparently stable Bode's law . . . configuration within a relatively short time; therefore Velikovsky's historical evidence cannot be ignored." By a relatively short time, Bass meant times of the order of centuries. He stated also that "three of the greatest contemporary mathematical celestial mechanicians have stated explicitly and recently that nothing known to them forbids Velikovsky's hypothesis."

Velikovsky's Later Work

Velikovsky contributed many articles to *Pensée* [418–20, 422, 423, 437–41, 443, 444, 446] and to *Kronos*. He published two sequels to *Ages in Chaos*—in 1977 *Peoples of the Sea* [413], and in 1978 *Ramses II and His Time* [414]. More of his work is being published posthumously; *Mankind in Amnesia* [415] appeared in 1982 and *Stargazers and Gravediggers* [416] in 1983. But perhaps his most significant piece after 1960 was his address ("My Challenge to Conventional Views in Science") at the symposium (in 1974) of the American Association for the Advancement of Science, where he reviewed and defended his work [445]:

> . . . what I offered is primarily a reconstruction of events in the historical past. . . . I did not set out to confront the existing views with a theory or hypothesis and to develop it into a competing system . . . all ancient civilizations . . . tell in various forms the very same narrative that the trained eye of a psychoanalyst could not but recognize as so many variants of the same theme. . . .
>
> Why is theomachy the central theme of all cosmogonical myths? . . . The legends and myths clearly point to an astral origin of all ancient religions. . . .
>
> Thus I entered a field that should be at the root of the natural sciences, not only of the human soul and of racial memories, and soon I observed that the divisions in science are artificial. I had to cross barriers. How could I do otherwise? Upon the realization that we are unaware of the most fateful events in human history, I had before me the task of explaining this well-known phenomenon of repression, the realization of which could also become crucial to the survival of the victim of amnesia playing with thermonuclear weapons. . . .
>
> . . . I was also carrying my heresy into a most sacred field, the holy of holies of science, to celestial mechanics. . . .
>
> The behavior of the scientific community was and partly still is a psychological phenomenon. The spectacle of the scientific establishment going through all the paces of self-degradation has nothing with which to compare in the past—though every time a new leaf in science was turned over there was a minor storm, and it is not without precedent that most authoritative voices in science should discourage the trail blazer—think of Lord Kelvin, unsurpassed authority of later Victorian days, who rejected Clerk Maxwell's electromagnetic theory, demeaned Guglielmo Marconi's radiotelegraphy, and

till his death in 1907 proclaimed Wilhelm Konrad Roentgen a charlatan.

But it is without precedent that the entire scientific community should be aroused to very base actions . . . against a solitary figure whose only iniquity was to present views carefully arrived at in more than a decade of work . . . with never a jest or a harsh word against those with whom the nonconformist disagreed. . . .

Now, after 24 years, . . . my *Worlds in Collision,* as well as *Earth in Upheaval,* do not require any revisions, whereas all books on terrestrial and celestial sciences of 1950 need complete rewriting. . . .

My work today is no longer heretical. Most of it is incorporated in textbooks and it does not matter whether credit is properly assigned. . . .

After all, it really does not matter so much what Velikovsky's role is in the scientific revolution that goes now across all fields from astronomy, with emphasis on charges, plasmas and fields, to zoology with its study of violence in man. But this symposium . . . is, I hope, a retarded recognition that by name-calling instead of testing, by jest instead of reading and meditating, nothing is achieved. None of my critics can erase the magnetosphere, nobody can stop the noises of Jupiter, nobody can cool off Venus, and nobody can change a single sentence in my books.

Velikovsky's Unpublished Works

Throughout the Velikovsky affair, Velikovsky and his supporters referred to books that remained unpublished for decades. "The publication of the second and final volume . . . is scheduled for a few months from now," wrote Velikovsky in February 1952 in the foreword to *Ages in Chaos.* In 1953 [301] "the second volume is set, however not yet published"; in 1955 "the second volume of Ages (spring 1956)," and "the second volume . . . is presently in galleys." That promised volume has become four promised books; in 1972 [292] "*Ramses II and His Time,* is scheduled to appear later this year . . . *Peoples of the Sea* . . . next spring." *Pensée* [297] explained that

> there has been no suppressive effort designed to prevent further publication of Velikovsky's books; the delays have been Velikovsky's own choice. The *Ages in Chaos* sequel, first set in type over 20 years ago as a single volume, has now grown to four or five volumes. Velikovsky has chosen to subject each volume to the most rigorous checking and cross-checking procedure possible. Two of these volumes are now [1973] near publication: *Ramses II and His Time* and *Peoples of the Sea;* the former may come out this summer and the latter shortly thereafter. None of the other works Velikovsky has been preparing, including *Mankind in Amnesia* and the story of the Deluge, are currently scheduled for publication.

In 1973 *Pensée* published [149] a chronology for the period from about 1000 B.C. to about 300 B.C., based on the unpublished sequels to *Ages in Chaos,* which were said to be *Assyrian Conquest* (830–612 B.C.), *Dark Ages of Greece, Ramses II and His Time* (612–525 B.C.), and *Peoples of the Sea* (525–332 B.C.). In 1974 publication of *Peoples of the Sea* was forecast [117] for the summer of that year. In the Winter 1974–75 issue of *Pensée* it was noted that "two [of these sequels] are in printers' proofs . . ." [447]. *Peoples of the Sea* was finally published in 1977, *Ramses II and His Time* in 1978.

In 1950 we read that "*Worlds in Collision* comprises only the last two acts of the cosmic drama. A few earlier acts . . . will be the subject of another volume of natural history" [408:preface]. Presumably, at least some of this material had been in the original manuscript; those who reviewed the latter for the publishers had "recommended the deletion of an earlier catastrophe" [166]. None of this natural history has yet been published, but there have been tantalizing references to some of the contents.

"In the near-collision with Jupiter, Saturn, of much larger mass then [*sic*] at present, was disrupted; the historical and folkloristic material on this subject fills a volume of the promised work on the earlier catastrophes, one of which was the Deluge . . ." [434]. The Deluge apparently [231] occurred in "the 24th/23rd centuries B.C. (end of the Old Kingdom in Egypt)—Velikovsky claims not to know this date with exactness." One gathers that "the catastrophe of the Deluge" is ascribed to "Saturn exploding as a nova . . . between five and ten thousand years ago" [438]; or "Saturn was fragmented by some kind of collision with Jupiter, and those fragmentary bodies which it did not reabsorb into itself caused disturbances through the rest of the solar system, both immediately and long after. The immediate disturbance affecting the earth was the deluge, caused by a watery cometary body of Saturnian origin. The principal later disturbance was caused by the instability of Jupiter, resulting from its absorption of Saturnian fragments. By a process of fissioning Jupiter generated the proto-planet Venus . . ." [265].

"In the sequel to *Worlds in Collision* I will discuss how Jupiter collected much of the dispersed material that resulted from the near-collision of major planets and then reached the point of disbalance . . ." [432]. Also, "the sequel to *Worlds in Collision* . . . deals with earlier catastrophes, at least one of which the human record ascribes to Mercury" [173]. That sequel was to be called *Saturn and the Flood* [347]; it had not appeared by 1982, although supposedly [134] written in the early 1940s.

At one time Velikovsky also envisaged publication of the full version of *Cosmos without Gravitation* [407]: an interviewer [35] had reported in 1950 that Velikovsky's "own explanation of the movement of the solar system . . . entitled 'The Orbit' . . . will probably be finished by the end of the summer." In the text of *Cosmos without Gravitation,* which is termed a synopsis, it is also stated that the full version is available in manuscript form. But that treatise on celestial mechanics has still not appeared in print.

Velikovsky talked of books on other subjects as well. For example, "a volume dealing with collective amnesia" was mentioned in 1972 [292]; it was published posthumously ten years later. Already in 1968 [233] he was "well along in a book called *Ten Trials,* which would set forth tests of his theories to be done by scientists to either prove or refute them." He "would also like to publish his correspondence with Albert Einstein," and "a book called *Ash,* a record of his struggle with the Scientific Establishment to have Egyptian material dated at radiocarbon laboratories. And then he has the idea that it might be worthwhile to publish some of the letters he has received from people around the world—say ten letters for each of the 18 years since the publication of *Worlds in Collision.* He has a prospective title for that book, too—*Letters to a Heretic.*" Other "manuscripts either in complete form or half done" include "several volumes on the early history of the controversy, and a volume on radiocarbon dating. (He's got hooked on carbon-14 dates, having gained some points on radiocarbon dates, but now having perhaps to confront radiocarbon dating antagonistically.) . . . There's . . . *The Test of Time,* a review of confirmations of his advance claims; a manuscript to do with the nature of Venus; *Gravediggers and Stargazers* [*sic*], a history of astronomical debate . . ." [349]. Velikovsky's account of his battles with scientists is entitled *Science and Conscience* [202, 254].

Velikovsky

Observers at the AAAS symposium in 1974 had given their impressions not only of the symposium but also of Velikovsky himself. Some were relatively sympathetic to the man, if not to his ideas [7, 48, 370], almost conciliatory: "One had to wonder what it was about the man that had inspired such outrage 24 years ago." One of the organizers of the symposium surmised that "all of us can see our own faults mirrored in his approach to science," and a panelist remarked that no one on the panel "believes Dr. Velikovsky is a crank in the usual sense, and some of us believe some of what he says is plausible" [117].

In turn, some of Velikovsky's supporters recognized weaknesses in his manner, if not in his ideas:

> This pattern of answers that grow increasingly rambling as time wears on—the 78-year-old Velikovsky had been forced to rest briefly in the middle of his prepared address—is one that by now has become familiar to Velikovsky-watchers. . . .

> Velikovsky himself ought to become aware that he is not always his own most effective defender. [304]

> As others before him, Velikovsky has pointed to the weaknesses of science, rather than the strengths, for here are where the discoveries and advances will be made. As a human being, he worries about his health, the troubles with his car, the sizeable telephone bill, and the mountainous correspondence on which he has scant hope of catching up. He is concerned about his books and papers yet to be published, and about the frustration of those who wait for him to provide new answers.
>
> He is very set in his ways, exasperatingly slow and deliberate, and, as even his friends will admit, is not easy to get along with. But, it is best to overlook any foibles of this man, as he is also a *rara avis,* a bennu-bird, that appears occasionally in the guise of a natural philosopher, attempting to shed a little more light on our ignorance. [166]

Immanuel Velikovsky died on 17 November 1979. Obituaries appeared in many places and continued to reflect the division of opinion about his work. In the *New York Times* [124] readers were told of "fanciful theories"; in *Industrial Research/Development* [171] Velikovsky was described as "one of the most controversial scientists of this century . . . acceptance of Velikovsky's work is inevitable."

Most significant, perhaps, was the following, because of the nature of the publication in which it appeared, the *Transactions of the American Geophysical Union* [27]: "Pioneer Venus results reportedly show Venus not to be the body predicted by Velikovsky . . . [but he] correctly predicted . . . polar wandering on Earth, the surface characteristics of Mars, RF emissions of Jupiter, and the high temperatures on the surface of Venus. . . . Velikovsky . . . had influence on the U.S. space program. . . . The growing interest in exploration of the planets during the 1970's was sparked and inspired by his thoughts and theories."

So opinions continue to differ about the merits of the case; indisputable, however, is that millions had been exposed to Velikovsky's ideas and that many had found welcome stimulation in them.

Something New, Something Old

In the 1950s the first violent battle had been fought: polemically, with sweeping generalizations ("genius" versus "crank"), and also with intricately complex arguments about the dates of eclipses, the provenance of references, etc. In the 1960s the matter of the advance claims had been central, and much was said about the behavior of the scientific establishment and the need to reform science. In that second engagement a group of supporters formed around Velikovsky; he gained moral support and entrée to campuses. At the same time echoes of the earlier barrages continued—intemperance on both sides, unwillingness to yield on the smallest detail, polemic and counter-rhetoric.

New in the 1970s was that Velikovsky's support became organized: societies, journals, symposia. New also was an air of civility on the part of some "establishment" scientists. But together with these novelties one found the mixture as before: polemic, maneuvering behind the scenes. Some scientists attempted to prevent the American Association for the Advancement of Science from holding its 1974 symposium about Velikovsky [117], and organizers of the symposium clearly displayed an anti-Velikovsky bias [318]. Publication of the proceedings of the symposium was disrupted [118, 169, 344, 450] by arguments about who could have how many pages, how many rebuttals could be printed, and how much time could be given to revisions; finally, the publication [119] contained only anti-Velikovskian contributions, and a response in book form was published by *Kronos* [201]. *Science News* failed to honor its contract [133] to carry advertisements for that book and for Ransom's book *The Age of Velikovsky. American Scientist, Sky and Telescope* [296], and *Scientific American* [308] had refused to accept advertisements for *Pensée.* The publisher of *Scientific American* wrote:

> We have not encountered a single scientist working in any of the many fields, from archaeology to astrophysics, on which Velikovsky touches who finds any interest whatever in anything he has to say. That is why you have not seen any account of Velikovsky in our pages. . . . The controversy seems to be generated wholly by Velikovsky and his sympathizers. They cry "foul" because he is ignored and attempt to make an academic freedom case of it. The controversy is thus quite secondary. As I see it, the threat to academic freedom comes the other way around: by such tactics the Velikovsky party tries to compel interest by scientists in work in which they can find no interest. [308]

One astronomer who was invited to participate in a Velikovskian symposium wrote, "Your suggestion that I should participate in a symposium of charlatanry is insulting. Please do not bother me any more with any of your material" [97].

And the Velikovskians responded by rehearsing their complaints. David Stove wrote about "The Scientific Mafia" [390]; Robert Treash, "A Candid Look at Scientific Misbehavior—Magnetic Remanence in Lunar Rocks" [403]; Lynn Rose, "The Censorship of Velikovsky's Interdisciplinary Synthesis" [341]; Horace Kallen, "Shapley, Velikovsky, and the Scientific Spirit" [183]; Euan MacKie, "A Challenge to the Integrity of Science?" [230]; Sidney Willhelm, "Velikovsky's Challenge to the Scientific Establishment" [464]. *Pensée* published 15 pages of correspondence dealing with Velikovsky's attempts to obtain C-14 dates of Egyptian objects [301]: "Velikovsky . . . had stated his expectation that . . . short-lived materials from Tutankhamon's tomb should yield dates ca. 840 B.C. The British Museum samples, tested seven years later . . . were dated to ca. 846 B.C. (BM642A) and ca. 899 B.C. (BM642B) respectively. . . . These dates, never published by the British Museum despite the assurance that they would find publication 'shortly', were set forth in . . . *Pensée.*"

G. W. van Oosterhout, having been unable to find these dates in the British Museum data in *Radiocarbon,* wrote to *Pensée* for further details. *Pensée* referred the matter to Bruce

Mainwaring (Foundation for Studies of Modern Science, Princeton, N.J.), who wrote to Oosterhout: "I have for some time been quite curious as to why these results had not been published. . . . Mr. Burleigh, the director of the laboratory of the British Museum . . . admitted that results which deviate substantially from what is expected are often discarded and never published. It is my personal opinion that that is what happened in this case." In the meantime Oosterhout had received a direct communication from the British Museum: "this laboratory has made no measurements on material from the tomb of Tutankhamun. . . ."

So the war continued to be fought quite vigorously. But the outcome remained tantalizingly inconclusive: the C-14 dates of Egyptian objects had seemed a possible way of arriving at a definite answer at least about Velikovsky's revised chronology, but that has turned out—at least up to the present—not to be the case.

A collection of articles from *Pensée* was published in book form under the title *Velikovsky Reconsidered* [315]. The reviews of that volume gave an opportunity to assess whether this latest battle had been in any way conclusive. Of the first 10 reviews that I saw, there were 3 pro- and 3 anti-Velikovsky, while 4 reserved judgment—not a notable change from 1950, when the first 32 reviews of *Worlds in Collision* were split 16 pro, 10 anti, and 6 uncommitted [82]. In fact, reading the comments on *Velikovsky Reconsidered* gave me a feeling of *déja vu*—all the themes were being played over again, and again, and again—the needle stuck in a groove, no resolution in sight.

On the one hand [123, 190, 374], readers of these reviews were told of Velikovsky's erudition, the scope of his research, his challenge to "accepted scientific dogma" [123], his successful advance claims, the boycott of Macmillan, the firing of Atwater and Putnam; that ". . . Velikovsky still awaits a less-prejudiced and better-balanced assessment by individuals prepared to use an unprejudiced approach" [123]; that "for over 26 years Velikovsky has been ruthlessly excluded from the columns of 'reputable' journals, even to enter defence against gross libel" [374].

More to the middle, there were comments like: "While many of Velikovsky's astronomical predictions have apparently been verified . . . , his theories . . . have been mostly ignored or distorted by the scientific establishment . . ." [327]. And toward the other side [1, 49, 277]: ". . . the bulk of the work is given to polemical innuendo, bad high-school physics, and disjointed proclamations of supposedly startling information deduced from unspecified assumptions. . . . obvious appeal to cultists . . ." [49]. Once again, the correspondence columns featured letters pro and letters con [50, 223, 227, 255].

A similar mixture can be found in response to *Peoples of the Sea* [170, 228, 394, 399]. After more than 30 years, one easily gained the impression that little had happened to clarify the issues, to provide clear and objective grounds for accepting Velikovsky's views or for rejecting them.

The Continuing Story

This chronicle of the Velikovsky affair was designed to set the stage for Parts II and III of this book: to make the reader familiar with the main themes of the controversy, with sufficient quotations and examples to convey the nuances and the clues on which I draw in my analysis.

In my view, no significant new phase of the controversy has clearly emerged since the early 1970s. Those who want to read the most up-to-date material will find much of it in *Kronos* and *S.I.S. Review*. Most pertinent, perhaps, is the continuing series of articles by Ellenberger [99], who cites and discusses material about Velikovsky as it appears in books and periodicals. Forrest [109] has compiled a lengthy comparison of Velikovsky's citations with the full original sources, arguing that Velikovsky's interpretations are anything but compelling. Explicit discussion of the Velikovsky affair was given space in *Zetetic Scholar* [79, 98, 241, 242, 258,

333, 475]; that journal also has continuing coverage of new articles and books germane to this and many similar subjects.

Inevitably, Velikovsky's death marked the beginning of a new stage of the Velikovskian strivings for recognition by the orthodox disciplines. There seems to be a splitting into separate sects of the once reasonably cohesive group of Velikovskians. *S.I.S. Review* publishes increasingly revisionist work [30, 158, 159, 186, 222, 371] as well as the more orthodox Velikovskian pieces, for example; and de Grazia [77] subscribes more to the value of Velikovsky's approach to conceptualizing than to the manifold specifics of Velikovsky's conclusions; whereas *Kronos* seems concerned to defend the validity of a broad range of Velikovsky's ideas and writings.

It is not inconceivable that there will be a resurgence of public interest. First, the subject of *Mankind in Amnesia* [415] is germane to matters of current concern—the determinants of human behavior, the dangers guaranteed by that behavior, the increasing number of sources of possible catastrophes. Second, unorthodoxies of various sorts now generate very widespread public interest, and attempts to debunk unorthodoxies are increasing [3, 32, 47, 69, 70, 114, 237, 329, 372]. Velikovsky's own account of the controversy, posthumously published [416], adds interesting details of the unprincipled conduct of some of his critics. The recent discovery [99A] of relevant Macmillan files and papers of James Putnam in the New York Public Library throws important new light on the dismissal of Putnam by Macmillan; on the manner in which Macmillan advertised *Worlds in Collision* [99B]; and on the advice Macmillan had received from those who read the manuscript before it was accepted and again just before it was published. But the analysis in the following chapters does not hinge on those details or on the future of the Velikovsky affair. Though I reach conclusions, my aim is to exemplify and stimulate a critical approach rather than to force assent with my personal convictions. A critical approach is most needed, after all, when questions remain open.

PART II

An Analysis of the Velikovsky Affair

Words are like leaves; and where they most abound,
Much fruit of sense beneath is rarely found.
—Alexander Pope

6

Is Velikovsky Right or Wrong?

> Now, who shall arbitrate?
> Ten men love what I hate
> Shun what I follow, slight what I receive;
> Ten who in ears and eyes
> Match me: we all surmise,
> They this thing, and I that: whom shall my soul believe?
>
> —Robert Browning

The question has been bandied about now for more than 30 years: some are certain that, yes, he is right; others are equally certain that he is wrong; others again leave the matter open still, or believe him to be right in part only. Where so many have disagreed for so long, can one reasonably hope to reach a definite conclusion?

I believe so. Not a definite answer to the question "right or wrong?" because that is too simplistic, quite meaningless in the context of something like the Velikovsky affair. But we can learn a great deal and reach some fairly definite conclusions by examining the nature of the question, and particularly what sort of answer might be possible. Two issues must be addressed before any meaningful answer can be given. First, *about precisely what* are we asking whether Velikovsky is right? Second, *with what degree of certainty* do we wish to be able to say "Yes" or "No"?

Not many who have debated about Velikovsky have analyzed the question, what exactly are Velikovsky's contentions? At first blush, the answer might seem obvious—that Venus came, as a comet, from Jupiter; caused catastrophes on earth about 1500 B.C.; and so on—Velikovsky's account of celestial events, in other words. But Velikovsky also contends that the Exodus occurred during a great natural catastrophe, one of the encounters of earth with Venus. If it turns out to be certainly proven that there was a catastrophe during Exodus, but one not caused by Venus, what would we then say? That Velikovsky was partly right? Or, if an encounter with Venus did occur but not during Exodus? What then? Velikovsky has made a whole host of statements on a wide variety of topics. Which of these are we to regard as crucial?

Thus no simple answer *can* exist to that initial, global question—unless one entertains the possibility that Velikovsky's contentions are all right or all wrong. I reject both those possibilities out of hand, and face a more difficult question: in how many details, and which ones, must Velikovsky be right in order to convince me that his cosmic scenario is likely to have a large element of truth in it?

Testing and checking Velikovsky's myriad inferences from historical and legendary sources will not be fruitful. In that respect *Worlds in Collision* and the *Ages in Chaos* series are as self-consistent as one could reasonably ask of such a massive compendium; the inferences may at times be far-fetched or dubious, but they are by no means illogical. Taken on its own terms, the work is plausible enough, and historians critical of it have not given chapter and verse to convince me that it is totally unsound. But I seek conviction, not an assurance of plausibility.

I might be convinced by Velikovsky if he had been able to predict accurately some significant phenomena that could not be explained on any other basis than his world-view and scenario. We have already seen that arguments based on his advance claims were inconclusive. Nevertheless, I want to reexamine that matter briefly for the sake of this personal analysis.

Venus Is Hot

Velikovsky was correct that Venus is hot. But the high temperature of Venus can also be explained in terms of the conventional wisdom; at the very least, no scientist has concluded that it cannot be. I do not expect a full explanation until the composition of the atmosphere of the planet is known more precisely, in particular what the clouds consist of—only then could one attempt a proper calculation of the heat balance.

Velikovsky's explanation was offered in a manner that is not conducive to testing. In order to test it, one would ask: with what energy was Venus expelled (or fissioned off) from Jupiter? What was its temperature then? What is its average heat capacity (how much heat does it absorb or lose for a given change in temperature)? By what mechanisms, and at what rate (based on those mechanisms), did it cool? If more heat was generated when it encountered the earth, how much? How was that heat distributed between earth and Venus? And so forth.

This is the argument of those who have said that Velikovsky's hypothesis cannot be tested because it is too vague, purely a qualitative and descriptive account. The argument does not prove Velikovsky wrong, but it does underline the fact that he has not presented a conclusive, testable chain of reasoning to support his predictions.

Jupiter Emits Radio Noises

Again, Velikovsky was right about this. However, he has not made clear how he reached this conclusion or how it follows from his reconstruction of events. Only when he does that can the significance of the successful prediction be gauged. As it stands, there is as much reason to class him as wrong as to class him as right in relation to Jupiter. Admittedly, there are radio signals, as he suggested—but he coupled the original suggestion with the acknowledgment that Jupiter is cold. Yet "by . . . 1962 the consensus was that the surface temperature . . . was about 1000°F" [233], not so different from that of Venus, described by Velikovsky as hot. Moreover, this very temperature had been estimated on the basis of the intensity and type of radio signals observed. Further, "in 1950 I'd have interpreted Velikovsky's predictions as claiming thermal radiation. . . . But Jupiter has no thermal spectrum" [313]. "Velikovsky's prediction was . . . useless in . . . its *lack* of detail—where to look in the radio spectrum (. . . it covers a factor of 10,000 to one in frequency); what to see there, that is the character of the source (Velikovsky didn't understand that two kinds of distinct non-thermal emission are produced) . . . I who am a specialist in the field am moved to ask myself, Did this physician writing in 1954 know more about the physics of radio emissions from planets than this astrophysicist 20 years later?" [310]. And again, Velikovsky failed to predict the existence of radio noises from Venus, yet the high temperature is calculated on the basis of those radio signals (see Chapter 4).

The whole field of radioastronomy is still young, and no radioastronomer would claim that all the relevant facts are in, or that satisfactory theory now exists to explain them. But that means not that existing orthodox theory is wrong, merely that it is incomplete.

Venus Is Anomalous

Venus rotates about its axis in a retrograde manner, the opposite sense of rotation to that of the other planets. This has been alleged to demonstrate that Venus came to be a planet in some different way from that in which the other planets were formed. But there is as yet no agreed mechanism for the formation of the solar system. Some form of condensation from a cloud of gas containing vortices seems to be plausible, but the implications of that idea have not been fully worked out. We do not yet know that the theories of the mainstream of science cannot account for Venus's retrograde spin. A special, *ad hoc,* explanation like Velikovsky's may not be necessary.

Further, in other ways Venus is *not* anomalous. For example, the densities of the planets are known.[1] All the terrestrial ones (Mars, earth, Venus, Mercury) have a density of

1. The information is given in some popular books on astronomy as well as in encyclopedias and the technical literature; for example, see Motz [262].

about 5 grams per cubic centimeter (between 4 and 5.8); the others (Saturn, Jupiter, Neptune, Uranus) have a much smaller density, between 0.7 and 1.6 grams per cubic centimeter. Apparently, then, the terrestrial planets were all formed in a similar fashion or at least from the same primordial material, and this was not the same material as that from which the outer planets were formed. If Venus came out of Jupiter, how does one explain these groupings by density?

Once again, we have not proved Velikovsky wrong but he has not shown that his explanation is the only possible one, or even that it is the most plausible.

For the present purpose, the lack of clarity about the correctness of Velikovsky's advance claims is not decisive or of crucial importance. Even if he had many significantly correct predictions to his credit, that would not logically constitute a proof of his main underlying theses. During the Velikovsky affair many individuals took the view that "no one could deny that if it [a new theory] makes enough predictions that are verified . . . then it must be regarded as plausible and admitted as such" [230]. On the other hand, Motz [260, 261] said, "verified predictions alone do not validate a theory"; "that correct deductions can be made from false premises is too obvious a truth to be argued."

That truth is not obvious to everyone, but illustrations of it abound. For instance, a famous theory in the development of chemistry was that of phlogiston. Phlogiston, the "material" of fire, was—logically enough—thought to be contained in every substance that could be burned; in the act of burning, phlogiston was released (or transferred to another substance). This concept made possible a great number of valid generalizations about chemical reactions, and predictions of new reactions that also turned out to work. The theory was "verified" by many chemists in many experiments. Nevertheless, the theory is now universally regarded as wrong, in part because the hypothetical substance phlogiston would have negative weight, an attribute as close to utterly impossible as one can get.

Phlogiston is merely one example of an incorrect basic premise that can lead to a host of interlocking and correct predictions. One cannot simply use the correctness of predictions as proof of the basic premise: that premise must also fit into our general scheme of understanding. Otherwise we reject it—for example, if substances having negative weight are called for. Successful predictions are presumptive proofs of the validity of a premise or theory only if the latter has some inherent prior probability in the light of existing knowledge. Velikovsky must establish the correctness of his scenario in some more direct manner since it is neither plausible nor probable in the light of existing knowledge. He must, if he wishes to be treated as a scientist, accept that the onus of proof for his views rests on him.

Consider the courses of action open to a man in Velikovsky's position. In 1940 he conceived the idea that the plagues of Egypt were actual physical occurrences, rather faithfully described in the Bible. Searching in various places, he came upon a translation of the Papyrus Ipuwer and was struck by how similar the events described there were to those in the biblical account. Could these be the same events? On the face of it, no—because the papyrus predated the Exodus by several hundred years. Could the dating of the papyrus be wrong? At this stage our imaginary scholar would have attempted to find some independent evidence to alter the dating of the papyrus. But Velikovsky did not do that. He *postulated* the different date and then proposed yet further, thoroughgoing revisions of accepted chronology.

Our imaginary scholar would have sought to publish an article arguing a more recent date for the papyrus. His idea would have been discussed by referees of his manuscript and, if it were then published, by other interested historians. A variety of views and approaches would have been employed to test the possible validity of the new proposition. But Velikovsky did not do this. He was subjectively convinced and took some further steps.

If the biblical accounts were of true events, in particular if there had been disturbance in the apparent motion of the sun accompanied by the fall of stones from the skies, then the event might have been an encounter of earth with a comet. Other parts of the earth than the Middle East would also have experienced unusual, possibly catastrophic events. Did they? In old records and in folklore and myths from around the world, he found descriptions that in some ways fit the same event.

Were those descriptions contemporaneous with the happenings in the Mediterranean? Our imaginary scholar would seek independent evidence for the dates of the described events, assemble that evidence, and again seek to publish it, to gain acceptance of this interpretation or at least to have it discussed. But, again, Velikovsky did not do this. He *assumed* that the similarities in the descriptions were compelling and decisive and that therefore the accounts must refer to the same time and to the same happenings.

Could legends be probed for valid details of such events after hundreds of years? Again, Velikovsky did not consider this question independently. Simply because the descriptions fitted his interpretation, he assumed that folklore could be sufficiently reliable for his purpose.

Perhaps the myths concerning gods and goddesses, whose names were those of the planets, described actual events in which those planets were involved? Our imaginary scholar would delve into this question. He would weigh different possible interpretations of these myths. He would compare the alleged planetary movements with those that would be concordant with our knowledge of planetary motions. He would be suspicious of any reported motions that were not consistent with current astronomical knowledge. To establish the possible correctness of the traditions, as opposed to the correctness of the astronomers' calculations, he would seek independent evidence. But Velikovsky, of course, did not do that. He *accepted* all myths and ancient records as long as they fitted his hypothesis; he ignored the conflict with astronomical knowledge and theory, and he ignored [139, 238] any myths that told other stories. His mind was made up as to the truth of his own notions, so he came to believe that the physical and astronomical theories that conflicted must be wrong.

Clearly, Velikovsky had not followed the approach that is the accepted one in scholarly investigations. He did not present for discussion his methodology, so that it could be judged apart from his conclusions and as a prerequisite to them; he used as a test of validity only concordance with his own premises; and then he asked that his conclusions be accepted despite their conflict with existing ideas in a variety of well-established intellectual disciplines.

It follows that Velikovsky ought not to have expected scholars in those disciplines to take his work seriously and examine it in the usual manner—he had not presented it in a form suitable for such examination. He did not invite criticism or discussion of the many steps he took, of the manifold details on which his case was based—he asked for acceptance of the whole thing, all the interconnections and conclusions. So, in effect, he asked that astronomers reject Newton because of a revised dating—to be taken on trust—of the Papyrus Ipuwer, without giving any reasons inherent in proven astronomical phenomena.

It was simply not reasonable for a psychoanalyst to ask, on the basis of unorthodox interpretations of historical and quasi-historical material, that astronomers change their physical theories. In any discipline the mainstream of methodology and theory exists because, on the whole, it seems to work adequately for the bulk of the available material in that discipline. Revolutions in the mainstream occur only when, within a given discipline, it becomes evident that too much is at variance with the accepted doctrines [204].

Now, the foregoing does not mean that Velikovsky is wrong. But it does show that Velikovsky could not be taken seriously by those who work professionally in the relevant disciplines. The material he presented was not testable, or even discussable, in the framework of those disciplines. It might be right, but *it could not be used.*

Altogether, then, it would have been reasonable for astronomers to react to *Worlds in Collision* in some such manner as: there is a large amount of material here that has nothing to do with astronomy and that we are not competent to judge; what is explicitly astronomical is partly speculation that cannot at present be tested, and partly seems to be at variance with what we presently accept on the very good grounds that it explains many things so well. The book gives no specific, quantifiable, proven astronomical data that can be used to test the suggested scenario. Velikovsky might be right, but we cannot do anything with his material. His work might have truth, but it is not scientific; it does not belong in the corpus of science. Perhaps the historians can do something with it.

That was not, of course, the astronomers' reaction. Their passionate denunciations and precipitate actions did nothing to help the public arrive at a reasoned evaluation of Velikovsky's work. I shall say more about the actual reactions later; for the present, consider what the situation might have been if the astronomers had reacted in fact as in reason they might well have done.

I put myself in the position of one who is asked by the prospective publishers to advise on this manuscript. Are the facts and references, as far as they can be checked, accurate and correct? Is the material sound in relation to the various disciplines touched upon? How plausible is it, that this will in time come to be judged the work of a scholar and not that of a crank? I would seek the opinions of experts in the various fields. With *Worlds in Collision,* it would then turn out that not many astronomers, biologists, geologists, historians, or physicists would have kind words to say. However, when pressed, they would have to admit that—though they might disagree with methodology and regard the inferences drawn as doubtful—they could not readily and decisively disprove the views put forward by Velikovsky.

It would be natural then to examine the prospective author's background and credentials. Had he written other things? How had they been received? In the present instance this approach would have yielded a rich harvest. It would have led to *Theses for the Reconstruction of Ancient History* [406] and to *Cosmos without Gravitation* [407]. The latter reveals Velikovsky to be ignorant of some of the most elementary ideas and approaches in physics and chemistry. Most strikingly, this ignorance is coupled with a readiness to discuss those subjects with an air of expertise. The author of *Cosmos without Gravitation* is an ignoramus masquerading as a sage.

That judgment will be supported in detail by an examination of *Cosmos without Gravitation* in the following chapter. Let me say first, however, that even if this judgment is correct, it would still not decisively prove that Velikovsky's scenario is wrong; it would merely make it very unlikely that I could take it seriously. When a man makes a fool of himself by discussing, at ponderous length, a subject of which he is ignorant, I tend to assume that he will act similarly in other fields and at other times—an unwarranted assumption but one that is commonly made. It is important to discuss that a little more, to examine the connection between subjective convictions and objective realities.

Probability and Certainty

Some statements convey a feeling of absolute certainty: "I shall, sooner or later, die," or "if a candle burns in an enclosed space, it will cease to burn when all the oxygen has been consumed." All experience gathered by man in the past and up to the present moment supports the validity of those statements. They are certainly true, provided that past experience is a valid base for extrapolating into the future. If not, however, then either of those statements could prove to be untrue in any particular case at any particular time. For example, if all physical events are determined by a supreme being who is wont to alter or suspend physical "laws" at will, then we cannot with certainty predict anything.

Scientific activity is based on assumptions, not clearly stated by scientists but much discussed by philosophers. In practice, scientists assume implicitly that there is some regularity, some orderliness, in nature; that similar situations will lead to similar results; that, if all relevant variables are controlled, experiments (or more generally experiences, phenomena) are in principle reproducible and that such reproducibility can be approximated in practice.

Scientific laws are shorthand statements that describe large numbers of experiments, experiences, by setting down what the common factors are, the "causes." Working with such laws, it is our experience that we can satisfactorily predict future events, "explain" past events, build machinery that works. That experience gives us confidence in the laws, and we tend to regard them, as time goes by, as absolutely true.

But that judgment of absolute truth is a valuation; it is a subjective assessment and it depends on the assumptions that we make at the outset, often (usually, in fact) without being aware that we are making assumptions. Belief in regularity, orderliness, reproducibility of the universe is common among scientists and more generally among mankind as a whole. It appears that human beings wish to be certain, about some things at least. In everyday conversation and thought we use the language of certitude—that is right, this is wrong, something is true. Our use of that language reflects the fact that we customarily also act on the basis of certitude, upon beliefs, upon the presumption that we can be certain about some things. Thus, once we have formed an opinion about something, we then tend to look upon the matter as a true-or-false one; we have translated our judgment of likelihood, of probability, into the shorthand of certainty and soon forget (if we were ever even aware of it) that we might have made a mistake in that translation.

Perhaps there are people who go through life making estimates of likelihood, and acting cheerfully even while knowing that their chances of success are, say, 70 percent or 10 percent. But surely it is more common to weigh a course of action and then to say, "I will do that," leaving out the ifs, ands, and buts. Having made the judgment that it is possible, we proceed to believe that it will be done—not that it can perhaps be, or that it will work 7 times out of 10 or once in 10, simply that it will.

My thesis, then, is that we all prefer to be certain than to be uncertain. We think and talk in the language of certainty rather than in that of probabilities. Yet the latter are the only objectively sustainable terms.

Returning now to "scientific" matters. Assuming that there is regularity and orderliness in the universe, some things can be predicted with certainty: death comes to all of us, candles need oxygen to burn. However, in such complex situations as the Velikovsky affair, we are dealing with matters where the laws are not sufficiently well established, or the detailed parameters sufficiently well known, as to make possible statements like that about the candle. We are trying to extrapolate beyond existing knowledge, and even assessments of probability cannot be made with total quantitative precision. We do not have laws and numbers that can be manipulated to give the result that Velikovsky is right with a probability of, say, 10 percent or 90 percent. We can only conclude, on the basis of individual judgments, that it seems to us relatively likely or relatively unlikely—for various reasons of our own, not necessarily shared by others. Our human penchant for certainty, however, ensures that sooner or later we will slip into the belief that Velikovsky is right or that he is wrong.

I attempted at first to find definite proofs in the criticisms made of Velikovsky's work and in the responses to those criticisms. Such proof was not forthcoming. In seeking to reach a personal judgment, I then tried to assess the plausibility of Velikovsky's scenario and concluded that Velikovsky had not built an objectively strong case, at least not a compellingly strong one—because he used as a test of his evidence only whether it accorded with his premise and did not adduce *independent* proofs for the validity of his interpretations of the various sources. He did not show that his interpretations were the only possible ones, or even that they were the most likely ones; he simply did not discuss possible alternatives. He did not present his case in the way scientists are expected to present a case, and so it is not a scientifically compelling one.

Nevertheless, I could not at that stage bring myself to reject Velikovsky's work entirely. In the past iconoclastic views presented in unorthodox ways have at times later turned out to be correct. Further, if Velikovsky were quite wrong, that would seem to put him clearly in the ranks of the cranks and pseudo-scientists, and he seemed too erudite for that. He had intellectual associations of long standing with such respected scientists as Hess and Einstein, and he was taken seriously by a number of respected humanists and social scientists.

Consequently, I had to search for other ways to reach a judgment. In particular, how sound is his competence in the fields that enter into his discussions—in plain language, does he know what he is talking about? In *Cosmos without Gravitation* I found clear proof that, in the physical sciences, Velikovsky does not know what he is talking about.

7

Velikovsky's Physical Science

That mine adversary had written a book . . .
—Job

Cosmos without Gravitation

In 1946 Velikovsky had published his ideas about the theory of gravitation and related matters in a monograph entitled *Cosmos without Gravitation,* in the series *Scripta Academica Hierosolymitana,* of which he was editor. This book was not cited in *Worlds in Collision,* and in the voluminous literature dealing with the Velikovsky affair it has been referred to only rarely. In the following, all quotations without a reference are from *Cosmos without Gravitation* [407].

"The fundamental theory . . . is: Gravitation is an electromagnetic phenomenon. . . . Electric attraction, repulsion, and electromagnetic circumduction govern . . . [the] movements [of planets and satellites]. The moon does not 'fall,' attracted to the earth . . . nor is the phenomenon of objects falling in the terrestrial atmosphere comparable with the 'falling effect' in the movement of the moon, a conjecture which is the basic element of the Newtonian theory of gravitation."

What exactly *is* the Newtonian theory of gravitation?

Newton's name is associated also with the laws of motion, of mechanics. The third law states that for every action there is an equal and opposite reaction. On that basis very precise calculations are made of what happens when bodies collide, and of how one can propel rockets by ejecting material in the direction opposite to that of the desired motion of the rocket, among many other applications. Newton's first law states that a body remains in whatever state of constant speed it happens to be in (including the possible state of rest, i.e. no motion) unless acted upon by a force; forces produce acceleration (or deceleration).

When a body falls through the air, having been initially at rest, there is evidently a force acting on it, producing an acceleration toward the ground. The moon circles the earth. In the absence of a force the moon would move in a straight line. Evidently, there is a force pulling it out of that straight line, toward the earth. It turns out that the acceleration experienced by the moon toward the earth is quantitatively the same as would be experienced by any body falling toward the ground at that distance from the earth. The simplest assumption, then, is that the forces are one and the same, and this force that attracts all material objects to the earth is called gravitation. Moreover, not just the earth but all material bodies exert gravitational attraction; the acceleration produced by a body is directly proportional to the mass of the body, to the amount of material that it contains. Small bodies exert a small attraction, large bodies exert a large attraction.

On this beautifully simple basis the movements of the members of the solar system are accurately calculable. Moreover, where discrepancies between observation and calculation were found, they have also been explainable on the same basis. For instance, certain irregularities were noticed in the orbit of Uranus. These were postulated to be perturbations resulting from the gravitational attraction exerted by an unknown planet. Calculations were made of the position (orbit) which that unknown planet must have to produce the perturbation, and thereupon the "unknown" was found by telescopic observation, in the predicted place. These steps were taken not once but twice, with success both times, leading to the discovery of Neptune in 1846 and of Pluto in 1930 (see, for example, Motz [168]).

So Newton's approach served to link together phenomena that had not previously been understood to be effects of the same force, gravity. Newton's approach has been used since he first propounded it, in innumerable different applications, with never an indication that it might be faulty. Now Velikovsky asks that this unified approach be abandoned, that we regard the fall of bodies in the atmosphere as a quite different type of phenomenon from that of the moon's attraction to the earth. Why should we? All our experience supports, quantitatively, the validity of the concept of gravitation.

Velikovsky asserts that gravitation is an electromagnetic phenomenon. Much thought and effort have been spent over many decades on the attempt to find a link between these two types of force. So far, no relationship has been found: the electromagnetic and gravitational properties of matter appear to be completely independent of one another. For Velikovsky to establish a plausible case for conjectures that are so contrary to virtually the whole accepted body of theory and observation, he would have to marshal powerful arguments indeed. What does he offer?

In *Cosmos without Gravitation* Velikovsky lists 25 "phenomena not in accord with the theory of gravitation." We will discuss them in more or less detail.

Atmospheric Composition

"Aside from several important facts . . . discussed at length in . . . WORLDS IN COLLISION . . . the following . . . are incompatible with the theory of gravitation:

"1. The ingredients of the air—oxygen, nitrogen, argon . . . are found in equal proportions at various levels of the atmosphere despite great differences in specific weights. . . . Why . . . do not the atmospheric gases separate and stay apart in accordance with the specific gravities?"

Velikovsky says that here is a simple, well-known phenomenon incompatible with the theory of gravity. If Velikovsky is right in this, one might well ask, why have scientists not been disturbed about it? Because, in fact, there is no incompatibility at all between these facts and the postulates of the theory of gravity.

There are three familiar states of matter: solids, liquids, and gases. In a solid or a liquid the individual particles (atoms or molecules) are held closely together by attractive forces (chemical bonds or electric-dipole forces or the so-called van der Waals forces), so that each particle has very little freedom of movement relative to its neighbors. For instance, when one end of a rigid rod is lifted, the particles in the middle of the rod, and toward the other end—all along the rod, in fact—also move upward (Fig. 1).

When a liquid is poured from one container into another, a stream of liquid flows. The shape of the liquid changes readily, because in liquids—by contrast to solids—the particles can slip and slide easily around one another. However, the forces of attraction between them still prevent the particles from flying apart: liquids do not spontaneously turn into fogs, sprays, or gases (Fig. 2)—they evaporate at only a small rate at temperatures below their boiling points.

The particles in a solid are tightly packed together, but they are not without motion. They vibrate continuously around their average positions, and on the average they are

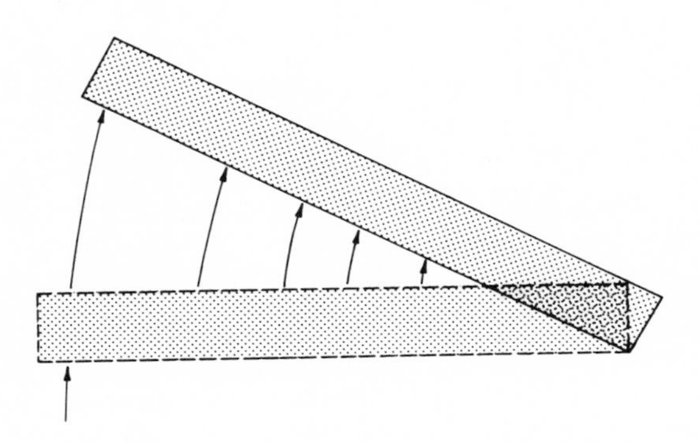

Figure 1. A solid rod is lifted at one end. The particles composing the rod all move upward although the lifting force is applied only to those at the end. The particles in a solid remain in fixed positions relative to one another.

Figure 2. Liquids, in contrast to solids, readily change shape. The particles in a liquid attract one another strongly enough that they remain in contact, but the particles freely move relative to one another.

arranged in a regular, ordered pattern (Fig. 3). Associated with the vibration is an energy of motion, kinetic energy. The magnitude of this energy is proportional to the temperature: the higher the temperature, the greater the kinetic energy, the more vigorously the particles vibrate. As the temperature rises, the motion of the particles tends to push them apart—the solid expands.[1]

As more and more heat is supplied, and the temperature of the solid increases, the particles vibrate more and more vigorously. Eventually, they have enough energy to leave their positions and to move around, interchanging positions with one another, no longer held in a regular ordered pattern—the solid melts, and forms a liquid (Fig. 4). A liquid

1. Different materials expand to varying extents for a given rise in temperature: they have different coefficients of expansion. We make use of that in practice: if the metal lid is "frozen" on a glass jar, we do not want to apply too much mechanical force (with a pair of pliers or a vise) for fear of breaking the jar. So we immerse the jar and lid in hot water—lo and behold, the lid can now be unscrewed. The metal has expanded more than has the glass (the coefficients of expansion of most metals are several times greater than those of common glasses; we understand why that is so, based on our knowledge of the forces between the individual particles of which those materials are composed).

is soft, not rigid, because the particles can slip and slide so easily around one another. Liquids take on the shape of any container into which they are put, whereas solids maintain their shape unless subjected to quite strong forces.

Although the particles in a liquid are free to move around one another, the energy of attraction between them is still larger than the average kinetic energy, so the particles do not fly apart from one another (Fig. 5). Liquids form pools or droplets, and we understand many of the properties of liquids (e.g. surface tension) because of our understanding of the attractive forces and of the variation of kinetic energy with temperature.

When liquids are heated, they evaporate—*some* of the liquid becomes a gas; at a sufficiently high temperature liquids boil—*all* the liquid is transformed into gas. Water becomes steam, for example. A given molecule in the liquid can leave it and become gaseous—can evaporate, in other words—if its kinetic energy is large enough to overcome the forces of attraction holding it to neighboring molecules in the liquid. The fraction of all molecules that has such a high kinetic energy increases with increasing temperature of the liquid, and at the boiling point the *average* kinetic energy is large enough to overcome the attractive forces (Fig. 6).

By the very definition of a gas, its molecules are in rapid motion because of their large kinetic energies. The particles move around freely in space, colliding with one another and rebounding, bouncing off the walls of the container. They also bounce off the ground, or off the bottom of the con-

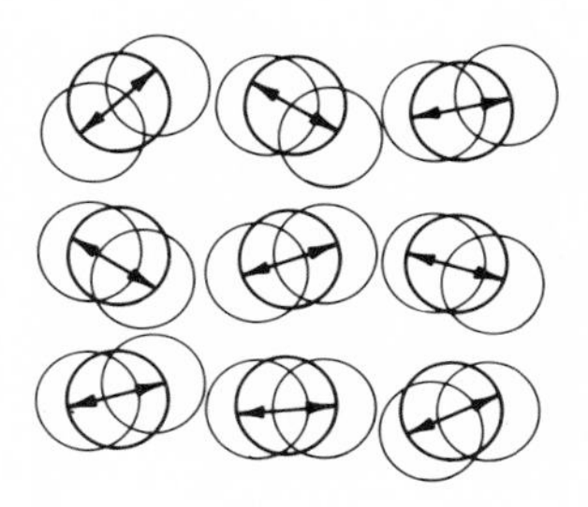

Figure 3. The particles in a solid are in continual vibration, but their *average* position does not change with time.

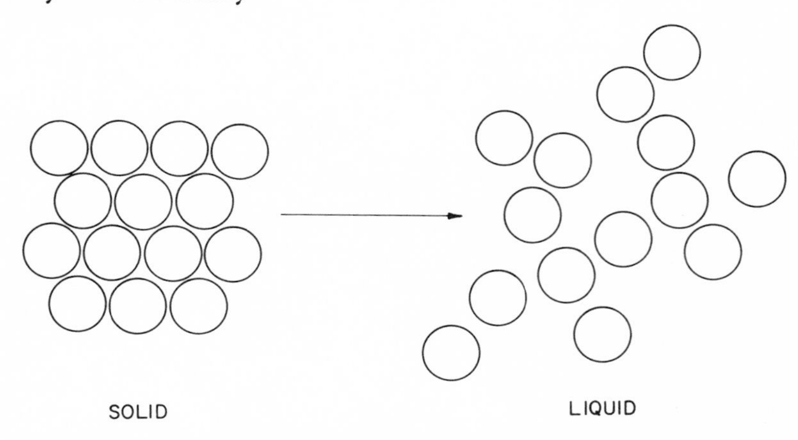

Figure 4. When a solid melts, the regular, ordered pattern of its particles is transformed into a random arrangement.

tainer, because their kinetic energies are greater not only than the forces of attraction between the particles but also than the force of gravity pulling the particles toward the ground (Fig. 7). Gases do not "settle down" under the influence of gravitational attraction; they fill completely any container into which they are placed. In a room we do not find all the air on the floor—it is distributed evenly throughout the room (Fig. 8).

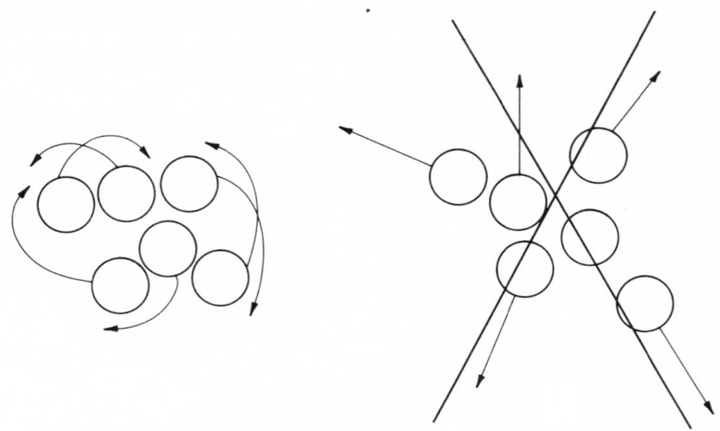

Figure 5. Particles in a liquid move easily around another but they do not fly apart from one another.

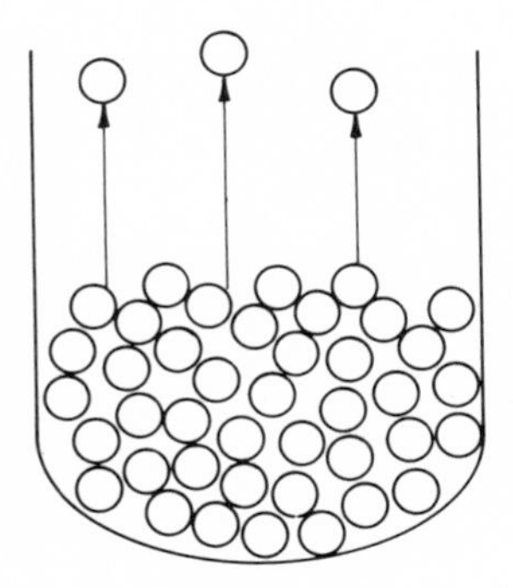

Figure 6. Particles leave a liquid when their kinetic energy is large enough to overcome the forces of attraction between them.

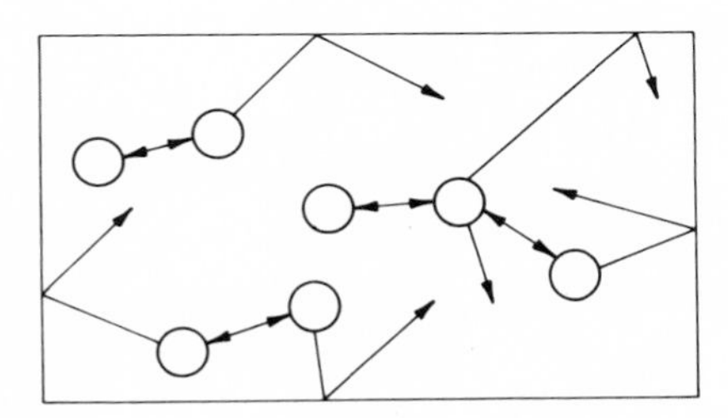

Figure 7. Particles of gas rebound from one another, the walls of the container, and the ground.

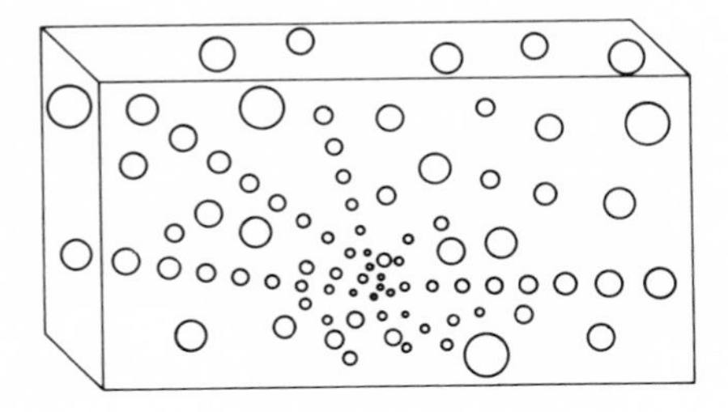

Figure 8. Gases fill completely the space available to them. The number of particles in a given small volume is, on the average, the same in different parts of the container.

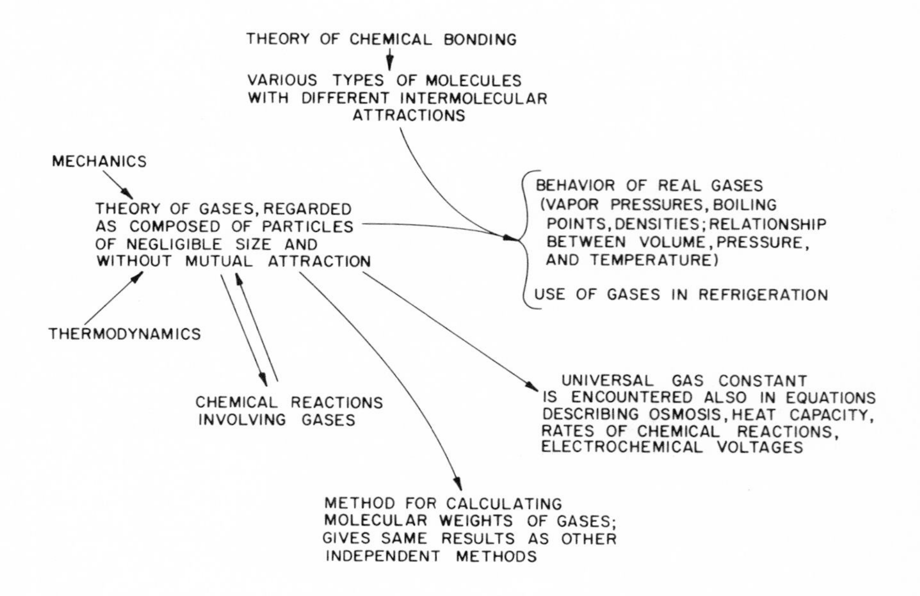

Figure 9.

This understanding of the behavior of gases is firmly linked to our understanding of many other natural phenomena (Fig. 9). By rather simply calculations, we can correlate the average kinetic energy with the temperature, the pressure, the density of a gas. The relationship among these leads to a remarkably useful parameter, the universal gas constant, which is encountered also in the quantitative description of many other phenomena—for instance, that of osmosis, the movement of liquid between various solutions (the use of "isotonic saline" in intravenous supply of glucose or drugs, for example, prevents damage to the red blood cells and is based on our understanding of the phenomenon of osmosis).

This discussion obviously has not done justice to the quantitative aspects of the kinetic theory of gases or to its manifold ramifications. Thus not all the particles in a given gas have the same energy—at a given temperature the *average* kinetic energy is fixed, but the energies of the individual particles are distributed over a range that straddles this average (Fig. 10).

Using the universal gas constant, one can calculate the molecular weight of any gas; or, for a known gas, one can calculate the number of molecules present in a given sample.

The relative magnitudes of the attractive forces between molecules of different gases can be understood in terms of prevailing concepts of chemical bonding and the distribution of electric charge within molecules. In turn, this understanding explains satisfactorily why different substances have different boiling points: for example, why that of mercury is very high and that of hydrogen very low; why the boiling points of the hydrocarbons increase regularly with increasing size of the molecules; and why hydrogen sulfide, though it is a larger molecule than water, has a lower boiling point. We understand why liquids boil at lower temperatures as the atmospheric pressure decreases—for example, the boiling point of water decreases as one proceeds to higher altitudes. Understanding of the behavior of gases and liquids is also used in air conditioning and in refrigeration systems, where cyclic compression and expansion—or liquefaction and evaporation—of gases is used to transfer heat.

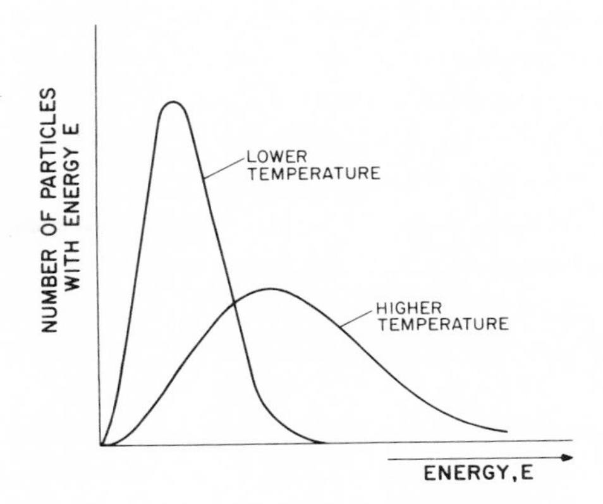

Figure 10. Not all particles have the same energy. Most of them have energies close to the average, but a few have much higher or much lower energies.

Altogether, then, the behavior of gases is integrally linked to many other phenomena, and our confidence in the validity of the theory, and of the concomitant equations, is justified by quantitative success in explaining and predicting literally innumerable experiments and experiences.

Now, this understanding of gases is not an arcane part of existing knowledge in the physical sciences. Most of the important features of the behavior of gases (e.g., that they do not settle out under the influence of gravity) have been known for centuries. The quantitative kinetic theory dates from the latter part of the nineteenth century. For many decades these matters have been routinely discussed in college-level courses in physical chemistry, quite commonly in freshman physics or chemistry, not infrequently in high school courses. It is surely reasonable to expect that one who purports to discuss the physical basis for the behavior of gases should be familiar with these concepts, their ramifications, their integral place in our knowledge of many phenomena in related fields. Yet Velikovsky reveals that he does not have such familiarity when he asks, "Why . . . do not the atmospheric gases separate and stay apart in accordance with the specific gravities?"

Consider the composition of the atmosphere at various heights. When two gases are present in the same container, the particles of each are in rapid motion and continually bounce off one another and off the walls of the container: the gases mix themselves. Each gas occupies the whole of the available space (even though most of it is empty space); we do not find two layers, with the "heavier" gas at the bottom, as we do with immiscible liquids, or when we place one solid on top of another. To a first approximation, we expect any gaseous mixture—including the atmosphere—to be uniform in composition throughout the whole volume that it occupies. The large kinetic energies of the molecules exceed the forces of attraction between them and the force of gravity—if that were not so, then the substances would not be gases, they would be solids or liquids.

However, in considering the earth's atmosphere, we are dealing with a "container" some tens of miles in height, and the first approximation of uniform composition will not be completely accurate. Although the kinetic energies of the molecules are large enough to prevent any of them from settling out, there should be some variation in average composition over large changes in altitude, with relatively more of the heavier constituents present at sea level and relatively more of the lighter ones at high altitudes. It is possible to make quite an exact calculation of the expected variation in composition with changing altitude, based on the theory of gases and that of gravity. Such a calculation leads to the following results [152:70]:

HEIGHT	*PERCENTAGE OF*			
(km)	*Nitrogen*	*Oxygen*	*Argon*	*All Other Gases*
30	84.26	15.18	0.35	0.21
20	81.24	18.10	0.59	0.07
15	79.52	19.66	0.77	0.05
11	78.02	20.99	0.94	0.05

Because the kinetic energies of the molecules are so great, the influence of gravity results in a very small effect only. The relative weights of nitrogen, oxygen, and argon are in the ratio 7:8:10, quite appreciably different—yet differences in altitude of many kilometers produce only slight changes in composition of the atmosphere.

In fact, measurements show that the atmospheric composition is even more constant, as the altitude changes, than these calculations predict. This extreme constancy is explainable on the basis of the extra mixing effects of winds, of temperature gradients, and of the slow rate at which the theoretically calculated "separation" shown in the table would come about after mixing ceases:

> . . . Swift winds keep the gases thoroughly mixed, so that except for water-vapor the composition of the atmosphere is

> the same throughout the troposphere [from sea level to about 10 km in height] to a high degree of approximation. . . .
>
> The effect of wind-mixing on the composition of the high atmosphere was first worked out by Maris. . . .
>
> . . . It was found that the proportions by volume of the respective gases were constant within a few per cent at all heights reached. The results were therefore in support of the foregoing diffusion-calculations. However, deviations from constancy, small, variable, and real were observed; these appeared to be attributable to irregularities in the lower atmosphere such as weather. [151:492–94]

In summary: on the basis of the kinetic theory of gases and the theory of gravitation, one expects a quite small change in composition of the air with changing altitude—not the gross separation that Velikovsky claimed should result from the effect of gravity.

Velikovsky answered his own query, as to why the atmospheric components do not separate, by stating: "The explanation accepted in science is this: 'Swift winds keep the gases thoroughly mixed, so that except for water-vapor the composition of the atmosphere is the same throughout the troposphere to a high degree of approximation.' This explanation cannot be true. If it were true, then the moment the wind subsides, the nitrogen should stream upward, and the oxygen should drop, preceded by the argon. . . ."

Here Velikovsky misrepresents the source [151:492–94] he is quoting: he makes it appear that it said that winds alone prevent the gases from separating, whereas the actual statement is that winds counteract, and prevent, the very small difference in composition with altitude that gravity would produce in the total absence of any mixing by winds. Moreover, the same source refers to the *slow* diffusion by which the differential composition tends to be set up after mixing ceases; yet Velikovsky states that the gases should "stream," "the moment the wind subsides." So Velikovsky gives a quite incorrect account of what "science" says, even though he has read—or at least quoted from—an authoritative source.

In the same place Velikovsky quotes a paragraph from Humphreys [152:70, Table II] about the nonoccurrence of pockets of foreign gas in the atmosphere. That book by Humphreys devotes many pages to a description of what is known about atmospheric composition and circulation, induced by gravitation and by temperature gradients. Velikovsky ignores all that and quotes a single paragraph which, taken completely out of context, could mislead readers into concluding that the authoritative references he cites support his case, when, in fact, they make the opposite point.

I conclude that Velikovsky is ignorant of the actual tenets of the theories of gravitation and of gases, despite the fact that he quotes books that correctly expound the theories and their implications. Either Velikovsky did not understand what he read, or else he deliberately misquoted those references. In either case what Velikovsky says about physical science is thereby suspect, even when he appears to quote chapter and verse from a respectable source. One had better check the source to determine whether it has been accurately or misleadingly cited.

Generalizing from a Single Case

I have spoken to only one of 25 "facts" said by Velikovsky to be incompatible with the theory of gravity. It is hardly legitimate to conclude that the shortcomings found there are necessarily to be found in the other 24. Must I then analyze each one in the same detail?

When I read *Cosmos without Gravitation* for the first time, I merely made mental note that, on this point, Velikovsky seemed to be ignorant of the kinetic theory of gases. Since I was not familiar with work on the composition of the atmosphere, I read some material about that. However, in attempting to make my line of thought clear to readers who might not themselves be familiar with the theories involved, it became necessary to write some six or seven pages of manuscript; to locate and read the relevant parts of two books quoted by Velikovsky (books containing a total of some 1,400 pages); and to revise my manuscript several times.

That illustrates, perhaps, why the scientists who reviewed *Worlds in Collision* did not attempt to rebut Velikovsky's physics and astronomy detail by detail. The labor involved would be simply enormous: not necessarily a difficult task but certainly a very time-consuming one, not worth the effort required (see Chapter 3, *The Case for Velikovsky*). So scientists said, "Velikovsky can be proved wrong," and left it at that.

Unfortunately, such a statement—no matter how sincere and even correct—is merely *ex cathedra,* or must at least seem so to a layman; it is a matter of Velikovsky's word against someone else's. I believe it important to avoid that unproductive type of confrontation, and wish to make the scientific argument accessible to all, so that the opportunity exists for everyone to reach an individually informed decision. So, having gone to some length with the first of the "facts" which, according to Velikovsky, contradict the law of gravitation, I shall proceed to show—albeit more succinctly—that similarly compelling arguments are available to contradict Velikovsky on each of the other 24 points.

The Other 24 "Facts"

"2. Ozone, though heavier than oxygen, is absent in the lower layers of the atmosphere, is present in the upper layers, and is not subject to the 'mixing effect of the wind.'. . . Nowhere is it asked why. . . ."

That question is not asked because it is a nonquestion; there is no paradox or anomaly here. Ozone *is* subject to the same convection as any other component of the atmosphere but, unlike most of the others, it happens to be unstable.

An ozone molecule consists of three atoms of oxygen chemically bound together. Ozone can be prepared from the normal, stable form of oxygen (in which there are two atoms of oxygen per molecule) by means of an electrical discharge. One frequently smells the sharp odor of ozone in the air after lightning has struck or in the output of electronic air cleaners or air fresheners under some conditions. The overall reaction involves three molecules of oxygen being transformed into two molecules of ozone:

$$\underset{\text{oxygen}}{3\,O_2} \longrightarrow \underset{\text{ozone}}{2\,O_3}$$

In the upper atmosphere this reaction occurs continuously under the influence of high-energy radiation impinging on the earth. That radiation is absorbed by the atmosphere (in part through the above reaction), and for that reason not enough of the radiation penetrates into the lower atmosphere to produce appreciable amounts of ozone below a certain altitude (though ozone is formed at lower altitudes, as already mentioned, locally and occasionally as in thunderstorms).

The ozone does tend to spread from the upper atmosphere, where it is formed, to lower altitudes, as a result of the normal self-mixing of gases and whatever convection arises from winds and thermal gradients. But ozone is quite unstable and spontaneously reverts into the normal form of oxygen. Thus ozone is present effectively only in the upper atmosphere because it is formed there and decomposes before it can reach the lower atmosphere. There is absolutely nothing here to cast doubt on the theories of gravity, of gases, or of the composition of and circulation in the atmosphere.

"3. Water . . . eight hundred times heavier than air, is held in droplets, by the millions of tons, miles above the ground. Clouds and mist are composed of droplets which defy gravitation."

There is a well-known class of systems in which the particles are larger than normal molecules but still sufficiently small that they do not settle out under the influence of gravity. These are called colloidal systems, made up of colloidal particles. Such colloidal particles are sufficiently small that the collision of molecules with them involves transfer of energy and momentum that is large compared to the influence of gravity, so that—like the molecules of a gas—they do not spontaneously settle out under gravity (Fig. 11). The motion of colloidal particles under buffeting by surrounding

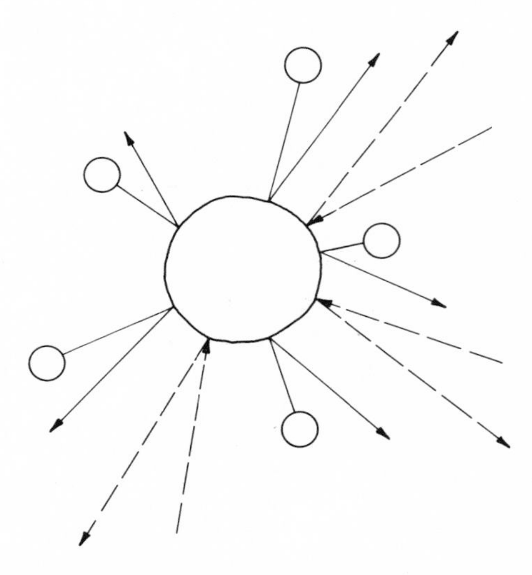

Figure 11. Particles of colloidal size (for example, water droplets in clouds) are continually buffeted from all sides by the fast-moving molecules of the surrounding medium. This buffeting prevents the particles from settling out under gravitational attraction.

molecules was first observed by a botanist, Robert Brown, in 1827. The explanation in terms of buffeting by molecules was arrived at during the last part of the nineteenth century, and a quantitative description of Brownian motion was given by Einstein in the early years of the twentieth century.

Colloidal systems are common in biology—such large "macromolecules" as proteins, enzymes, and nucleic acids behave like colloidal particles. Emulsions are another common example of colloidal systems (for example, the water-based "plastic" paints). All textbooks of physical chemistry give explanations and examples of colloidal behavior. Substances in the colloidal state can be made to settle out by causing the colloidal particles to aggregate into larger ones, until they become so large and heavy that the attraction exerted by gravity on an individual particle outweighs the force imparted by buffeting by surrounding molecules (which produces a random movement in all directions).

Clouds, mist, and fog consist of particles of water that are of colloidal size. These droplets remain suspended in the air as long as they do not exceed a certain critical size. One can induce the droplets to coalesce and aggregate—for instance, by "seeding" with crystals of silver iodide, which provide nuclei upon which large drops can grow: this is the basis for producing rain artificially.

This "fact" of Velikovsky's, then, serves only to show that he is ignorant of an important and well-understood class of substances. There is nothing here that is incompatible with the law of gravity. The influence of gravity on colloidal particles is simply not large enough to make them settle out (though colloids can be made to settle out by applying artificially high "gravity," as in a centrifuge—the same type of device as is used to accustom astronauts to the high accelerations encountered in the take-off of rockets).

"4. . . . the motion of the molecules, if effected by a mechanical cause, must subside because of the gravitational attraction between the particles and also because of the gravitational pull of the earth . . . as the molecules of a gas at a constant temperature . . . do not stop moving, it is obvious that a force generated in collisions drives them. . . ."

Again, Velikovsky demonstrates his lack of understanding of what gases are: by definition, substances in a state in which the molecules have such a large energy of motion that the attraction of gravity is virtually negligible in comparison.

At a fixed temperature the average kinetic energy of the molecules is a fixed quantity; the kinetic energy reflects, indeed is a measure of, the temperature (and, of course, vice versa). The molecules will stop moving, or slow down, or "subside," only if energy is removed from the system; the removal of energy from the system would be accompanied by a decrease in temperature—it is the same as a decrease in temperature. When two molecules collide, energy may be transferred from one to the other, but it is not lost; thus the *average* energy of all the molecules remains the same. When molecules collide with the walls of their container, they may transfer energy to the walls; if, however, the walls and the gas are at the same temperature—which is inevitably implied by the statement of "a gas at constant temperature"—then *on the average* molecules colliding with the walls receive as much energy as they lose, and there is no reason for the molecules to subside.

Once again, Velikovsky's use of references invites criticism. He quotes here from "Sir James Jeans, The Kinetic Energy of Gases, 1940": "There should be also a loss of momentum as the result of the transformation of a part of the energy of motion into vibration of molecules hit in the collisions. . . ." This leads the reader to believe that Jeans agrees with Velikovsky that the molecules should subside. The very opposite is the case. In the first place Jeans gives a full account of the kinetic theory of gases, quite similar to that given earlier in this chapter, and makes clear why the influence of gravity does not cause the atmospheric gases to settle out.

> An average molecule of ordinary air moves at about 500 meters a second, so that the parabola which it describes under gravity has a radius of curvature of about 25 kilometres at its vertex. . . . This is so large . . . that we may, without appreciable error, think of the molecules as moving in straight lines at uniform speeds, except when they encounter either other molecules or the walls of the containing vessel. This . . . explains why a gas spreads immediately throughout any empty space . . . there is no need to suppose, as was at one time done, that this expansive property is evidence of repulsive forces between the molecules. . . . [161:7–9]

Rather explicitly here, Jeans states that there is no need to invoke repulsive forces or, as Velikovsky put it, "forces generated in the collisions."

In the second place Jeans discusses [161:7–9] Brownian motion, of which Velikovsky is apparently ignorant when he states that the droplets of clouds and mist defy gravity. Finally, what Jeans actually says concerning "loss of momentum" is: "how is it that a gas . . . does not . . . lose the energy of translational motion . . . and replace it by energy of internal vibrations . . . and of radiation . . . ? . . . it was only after the introduction of the quantum theory . . . that it became possible to give anything like a satisfactory explanation. . . . The newer system of laws, constituting the modern theory of quantum mechanics, is beyond the scope of the present book" [161:15–16]. In other words, the "anomaly" to which Velikovsky refers, and for which he cites Jeans as authority, is explicitly stated by Jeans to be not an anomaly but a situation that is explicable by means of quantum mechanics.

"5. . . . The semidiurnal changes in barometric pressure are not explainable by the mechanistic principles of gravitation and the heat effect of solar radiation. . . ."

Velikovsky quotes Lord Rayleigh (1890) and Humphreys (1940) to the effect that these changes are not understood. That does not, however, make them not explainable, not understandable, on the basis of presently accepted physical laws, and Rayleigh and Humphreys do not say that—merely that a full explanation has not yet been given.

Innumerable phenomena have not yet been explained, particularly not in quantitative detail, but that is no argument for rejecting laws that work satisfactorily wherever it has been possible to apply them. Before rejecting those laws, it would be necessary to show that no conceivable combination of known influences could, under those laws, produce the phenomena in question. It would be further necessary to show how some proposed new laws could, while at the same time also accounting for all the phenomena that are satisfactorily accounted for by the "old" concepts. The argument used by Velikovsky on this point is typical of arguments that he frequently makes. He points to something that is not yet fully explained and then claims this as evidence that it is not explain*able* (or says "unexplainable" when he ought to say "unexplained"); therefore Newton, or Einstein, or Darwin, for example, was wrong. But there is no such "scientific establishment" that claims everything to be already understood. To overturn existing theory, one must adduce facts that contradict that theory, not merely point to facts that have not yet yielded to detailed explication.

The same objection applies to Velikovsky's "facts" 6 through 25. It is also worth noting throughout that many of Velikovsky's references are to early, not to say outdated work, going as far back as 1855. In scientific discussions one expects to see also the most recent work quoted, so that one can adequately judge the present state of understanding of the subject.

"25. The earth is a huge magnet. . . . As the principle of gravitation leaves no room for the participation of other forces in the ordinary movements of the celestial mechanism, these obvious and permanent influences of the electromagnetic state . . . are not allowed to have more than zero effect on the astronomical position of the earth. . . ."

Once again Velikovsky misrepresents the prevailing view held by scientists. The principle of gravitation certainly "leaves room" for the participation of other forces. Thus Bailey [20] was approvingly quoted by Velikovsky and his supporters for the suggestion that the sun carries a large electric charge, but Bailey, a physicist, had no difficulty in conceiving, and making calculations about, simultaneous electrical and gravitational effects. Because the observed motions of the planets and of their satellites do not differ from those calculated on the basis of gravitational effects alone, astronomers were simply not impressed by Velikovsky's insistence that electromagnetic forces should somehow be brought into consideration. Forces other than gravitation evidently exert an effect that is negligible compared to that of gravity and consequently they are not taken into consideration.

The misleading argument given here by Velikovsky is again typical of much of his writing. He does not distinguish between a situation in which some influence is calculably or observationally so small that it can be neglected, and a situation in which a principle, law, or force is not considered at all. Yet the making of precisely such distinctions is crucial to and fundamental in scientific activity. For instance, the influence of gravity on the composition of the atmosphere is normally ignored, not because gravity is in some way incompatible with the behavior of the gases or contradicts formulations of the kinetic theory, but because the influence of gravity is known to be so small that it affects the composition only when differences of tens of kilometers in altitude are considered.

The neglecting of influences that are known to be small makes scientific activity efficient. Rather than to perform horrendously complicated calculations, at the end of which some influences are seen to be negligible, one uses experience and makes the calculations in the first place considering only what are known to be important effects. Then, if the results happen not to fit observation, other effects are taken into account—only when and as needed.

Another example of Velikovsky misleading in this manner occurs when he quotes "H. N. Russell" and comments upon the quote: "'An atom differs from the solar system by the fact that it is not gravitation that makes the electrons go round the nucleus, but electricity.' . . . Different principles are supposed to govern the motion of the planetary bodies in the macrocosm and microcosm . . ." [408:387–89]. Scientists do *not* suppose that "different principles" are involved. Within atoms, electrical forces outweigh gravitational ones to the extent that the latter can be neglected for the purpose of even very exact calculations, and vice versa in the case of the solar system.

One might note also that Velikovsky himself is quite prepared to invoke two principles instead of one when it suits his purpose: he does not admit that the attraction to the earth of the moon is the same type of phenomenon as the attraction to the earth of bodies falling through the atmosphere.

The New Theory

Having, in his view, demolished the theory of gravitation, Velikovsky gives "general suggestions" toward a new theory: "1. Attraction between two neutral atoms. . . ." He appears to believe that such attractions are presently ascribed to gravity, which is not the case. Gravitational attraction between atoms is negligible compared to the attraction exerted by the van der Waals (or London, or "exchange") forces, whose mode of action is, in fact, describable qualitatively in terms quite similar to those used by Velikovsky in making this "suggestion." He is apparently ignorant of the existing description of these forces, although the relevant work predates his suggestion by a decade or two.

Another surprising statement by Velikovsky here is: "The solar surface is charged negatively . . . as the spectral lines (with the dominant red line in the spectrum of hydrogen) reveal. . . ." The spectrum in fact tells us nothing about the existence of a charge on the solar surface; if it did, Velikovsky's idea that the sun carries an appreciable charge would have been the commonly accepted one long before Velikovsky.

Finally, Velikovsky mentions 15 phenomena that can be explained by his new theory. His explanations fail to carry any conviction; it is not even clear from them in what way they are supposed to explain the phenomena. For example,

> 2. *The planets have a greater aggregate energy of motion than the sun.* The revolution of the planets did not originate in the angular velocity of rotation of the sun; the magnetic field of the sun effected their revolution. Also, the fact that one of the satellites of Mars revolves with an angular speed greater than that of the rotation of this planet is explained here by electromagnetic circumduction. . . .
>
> 10. *The influence of the moon on radio reception.* The charged moon on its hourly stations exerts an attracting-repelling action on the electrified layers of the atmosphere (ionosphere) in a greater degree than on the "insulating layer" of the earth's atmosphere. . . .
>
> 15. *Increased gravity over the sea.* The increase of gravity over the sea as compared with that over the continent may be explained by the higher charge of salt water.

Salt water, however, carries no "charge"; neither does pure water.

Summing Up

Cosmos without Gravitation shows Velikovsky to be out of his depth when it comes to physical science. He misrepresents theories and facts extant at the time he wrote—and not because he had not read about them, for he quotes from authoritative sources: either he did not understand, or he deliberately misrepresented. Whatever the reason, it is plain that Velikovsky's references to scientific work are not to be relied on. He uses the jargon of science as though he understands it. His tone is that of one who is discussing subjects with which he is familiar, yet a close look at what he says reveals that he is not competent to carry on such discussions.

Taking *Cosmos without Gravitation* by itself, I would not hesitate to characterize the author as a crank or a charlatan. Yet *Cosmos without Gravitation* is not the whole story. What of Velikovsky's historical works? Perhaps he is competent in that area? And if he is a crank, how to explain the fact that many highly intelligent and educated people take him seriously? Perhaps *Cosmos without Gravitation* was only an early aberration?

Worlds in Collision and Later Work

Cosmos without Gravitation is not cited in *Worlds in Collision*. Had Velikovsky perhaps recanted his earlier, mistaken criticisms of gravitational theory? It seems not:

> . . . The accepted celestial mechanics, notwithstanding the many calculations that have been carried out to many decimal places, or verified by celestial motions, stands only *if* the sun, the source of light, warmth, and other radiation produced by fusion and fission of atoms, *is as a whole an electrically neutral body,* and also if the planets, in their usual orbits, are neutral bodies.
>
> Fundamental principles in celestial mechanics including the law of gravitation, must come into question if the sun possesses a charge sufficient to influence the planets in their orbits or the comets in theirs. In the Newtonian celestial mechanics, based on the theory of gravitation, electricity and magnetism play no role. [408:387–89]

So Velikovsky still held the views expressed in *Cosmos without Gravitation* (as is also clear from his memoirs [416:165n]): he merely chose not to press his attack on conventional science along that particular line at that particular time. Rather, he stressed his historical reconstruction and, since then, the supposed success of his advance claims; the lack of receptivity on the part of the scientific community to his new ideas; the failure to analyze his work objectively; the dogmatic adherence to Newton, Darwin, and others, to uniformitarianism as opposed to catastrophism; and the lacunae and imperfections in existing theories and explanations. Yet behind all this stands the little known and rarely referred to *Cosmos without Gravitation,* which reveals Velikovsky himself to be an arch-dogmatist who regards his own intuition as a more valid guide than all the accumulated body of fact and theory in the physical sciences.

There are many other points in *Worlds in Collision* and later writings that reveal the same ignorance of the physical sciences as was shown in *Cosmos without Gravitation,* albeit not so strikingly, because the chemistry and physics enter only in a by-the-way fashion. Some examples follow.

The Atom and the Solar System

Velikovsky sees the atom as a valid analogy to the solar system and believes that he is bringing modern physics into astronomy by considering electrical effects and changes of orbit [408:preface, 387–89] (see Chapter 2, especially *Summing Up*). He is quite wrong. Electrical effects are not included in astronomers' calculations because accurate descriptions and predictions result from calculations based solely on gravitational effects. On the other hand, gravitational attraction between the positively charged nucleus of an atom and the "orbiting," negatively charged electrons is not considered by physicists because, although such attraction is certainly believed to exist, its magnitude is negligibly small compared to that of the electrical force. Moreover, electrons are not comparable to, or analogous with, planets. Not since the first quarter of this century have scientists regarded electrons as charged particles orbiting around the nucleus. Unfortunately and for historical reasons, we continue to use the term "electron," with its implication of particlelike character. We do, in certain circumstances, speak of electrons as though they were particles and sometimes use equations valid for particles to make certain calculations. At other times we use equations for wave motions to describe certain properties of electrons. The latter are something for which no commonsense, everyday description fits the facts, something that has both wavelike and particlelike properties. By contrast, there is nothing in the behavior of planets that can—let alone needs to—be described by wavelike equations or calculations.

"Changes in electron orbits"—more accurately, changes in energy levels of the electrons—do occur by absorption or emission of photons. More accurately yet, atoms and molecules absorb or emit electromagnetic radiation only by certain definite, discrete amounts of energy, called photons, whereby the energies associated with the electrons in those atoms and molecules also change by such discrete amounts. In such interactions the discreteness has to do with Planck's constant, whose magnitude is about 6.6×10^{-27} erg-sec: a truly tiny amount, significant only at the atomic and subatomic level. Where matter in macroscopic amounts is concerned, energy can be absorbed or emitted effectively in any arbitrary amounts; no discreteness effects are observed, and the concept (or description) of "photon" is meaningless. (More precisely, the magnitudes of the discrete increments of energy are so small compared to the overall changes in energy observable at the macroscopic level that no sign of discreteness is evident.) Planets, but not electrons, can gain or lose energy in a continuous (nondiscrete) manner, and any analogy between "orbital changes" in atoms and in the solar system is fundamentally misleading.

On occasion, supporters of Velikovsky have attempted to play down his comparison of atom and solar system [212]. Velikovsky's own words, however, show that he is serious about the analogy and believes it to be valid (Chapter 2, especially *Summing Up*).

Electrical Phenomena

"When the tidal waves rose to their highest point . . . a tremendous spark flew between the earth and the globe of the comet, which instantly pushed down the miles-high billows . . ." [408:77]. Nothing in what is known of electricity and matter indicates that an electrical spark would push down a wave, let alone instantly, let alone "miles-high billows." This sort of description, to be found throughout Velikovsky's work, the scientific community calls "hand waving": an explanation so vague that it cannot even be discussed, let alone disproved or proved. What current at what voltage for how long is needed to push down how much water, by what mechanism? On what law is such an assertion based? Without answers to those questions we have only hand waving, not science, and Velikovsky does not provide even clues to the answers.

". . . the tail of the comet and its head, having become entangled with each other by their close contact with the earth, exchanged violent discharges of electricity. . . . In the exchange of electrical potentials, the tail and the head were attracted one to the other and repelled one from the other . . ." [408:77]. This is a collection of genuinely meaningless statements. How, for example, do two entities "exchange . . . electrical potentials"? When charge moves under the influence of a difference in electrical potential, it moves always in the same direction, and the movement of charge acts to decrease the difference of potential. "Exchange of potentials" is literally without meaning: connect two poles of a battery, and it discharges—the poles do not "exchange potentials." Opposite charges (positive and negative) attract one another; like charges (positive and positive or negative and negative) repel one another. How could two bodies alternately, or simultaneously, attract and repel? It is not even clear from Velikovsky's description whether simultaneously or alternately is intended. This is once again mere hand waving, or worse—from a technical standpoint it is gibberish.

". . . interplanetary electric discharges could also initiate atomic fissions with ensuing radioactivity and emission of heat" [408:366–71]. Atomic fissions initiated by such electric discharges would be a new phenomenon. Enormous amounts of energy per nucleus are needed to produce such effects in the machines built to study atomic fission (cyclotrons, bevatrons, nuclear accelerators, etc.); the yields in these experiments are frequently no more than a few atoms of the new material. Once again, this suggestion of Velikovsky's is nothing but hand waving.

Chemistry

Velikovsky uses the words of chemistry in a manner that might impress a layman; to a chemist, however, this is just more waving of hands. According to Velikovsky, earth (and also moon and Mars) acquired from comet-Venus such substances as manna and petroleum, which are compounds based on carbon. Velikovsky talks of "the carbon clouds that precipitated honey-frost" [408:137]; "carbonigenous clouds in which the earth was enveloped" [408:171]; "white precipitated masses on Mars . . . polar caps . . . are probably of the nature of carbon"; "the envelope of carbon clouds of the Morning Star . . . clouds of dust"; "hydrocarbon gases in [Venus's] atmosphere" and "hydrogen and gaseous carbon" [408:366–71]. Apparently indiscriminately and interchangeably, there is talk here of "carbon"—meaning the element, in its elemental form; of "carbonigenous" and "of the nature of carbon"—chemically without meaning because of their lack of specificity; of "hydrocarbon"—referring to a definite type of compound that contains only carbon and hydrogen. There is nothing here that a chemist can use, can attempt to think about in a meaningful way, can assess as to plausibility; it is simply not clear what Velikovsky is saying. Even though the *language* of chemistry is used, the *meaning* is not chemistry.

". . . It takes millions of years for a log to be turned into coal but only a single hour when burning . . ." [411:preface]. No further explanation is given. Now, when wood is burned—incompletely burned, that is—it can be converted into charcoal, which is a very different substance from coal (the latter, by the way, is not a single, well-defined type of substance—one should speak of "coals" or "a coal"). Does Velikovsky not know that coals and charcoal are entirely different materials?

About the moon, Velikovsky [417] writes that he "would not be surprised if bitumens (asphalts, tar or waxes) or carbides or carbonates are found"; if any were found, no doubt that would become one of Velikovsky's successful advance claims. But neither would any chemist be surprised if those were found, though no chemist would think it was worth making the remark, because those are simply the most common forms in which carbon occurs. One does not need Velikovsky, his theory, or his historical reconstruction to make such a statement—or the prediction that "chlorine, sulphur and iron in various compounds, possibly oxidized, will be found richly presented in lunar formation . . .": those three are quite abundant, common elements.

Velikovsky's suggestion that petroleum came to earth as a result of an encounter with Venus is also couched in such vague terms that virtually any interpretation could later be said to have been the author's intention:

> If the petroleum that poured down on the earth . . . was formed by . . . electrical discharges from hydrogen and gaseous carbon, Venus must still have petroleum. . . .
>
> . . . If . . . Venus has petroleum gases, then Jupiter must have petroleum. . . .
>
> . . . Venus and Jupiter must possess an organic source of petroleum. . . . [408:366–71]

These statements are not even self-consistent. Is Velikovsky saying that the petroleum is of inorganic origin (formed electrically from carbon and hydrogen), or is he saying that it is of organic origin? His chain of reasoning as to its occurrence on Venus and Jupiter depends on which of these he means, yet he mixes them together.

Velikovsky refers to "the poisonous gases methane and ammonia" [408:366–71]; neither of those substances is a poisonous gas. Both are gases that can cause asphyxiation, as can all other gases than oxygen, but they are not poisonous or dangerous in the way that chlorine, or carbon monoxide, or hydrogen cyanide, or many other gases are. Methane is the chief constituent of natural gas; ammonia is readily available in a number of commercial cleaning solutions. No truly poisonous substance is so widely distributed and used in this poison-conscious society.

One little argument in 1950 concerned Velikovsky's statement that "absorption lines of argon and neon have not yet been investigated . . ." [408:366–71]. Haldane [139] commented sarcastically that this statement may have been intended to warn chemists and physicists that the work was a hoax—an appropriate sarcasm because the absorption lines (the wavelengths or "colors" at which light is absorbed by these gases) were accurately known from long before. In replying to this, Velikovsky [428] quoted a letter to himself from a prominent astronomer, to the effect that "line-spectra of these gases are well known but, so far as I know, their band spectra [lines of absorption] have never been studied." It appears from Velikovsky's insertion, "[lines of absorption]," that he had asked the astronomer about band spectra, believing band spectra and lines of absorption to be synonymous.

In fact, line spectra are characteristic of matter in the form of atoms, irrespective of whether light is being absorbed or emitted, whereas band spectra are characteristic of matter in the form of molecules (two or more atoms chemically bound together). Argon and neon, so-called inert gases, do not form molecules under any conditions so far known. That astronomer must have been quite puzzled when Velikovsky asked him about band spectra of argon and neon. What Velikovsky wanted to know, no doubt, was whether *absorption lines* characteristic of argon or neon had been found in planetary spectra, but he knew so little about the subject that he could not phrase his question correctly. The exchange serves to underscore Velikovsky's ignorance of the elementary facts of spectroscopy. As Haldane pointed out, "When neon or argon is hot enough or electrically excited, each gives off light of certain frequencies, and absorbs light of the same frequencies when a stronger light shines through it. . . . You can observe it in any London tube train with a pocket spedroscope [*sic*]" [140]. To say, in the 1940s, that absorption spectra of argon or neon were not known was simply to display ignorance.

More recently, Velikovsky raised "a fundamental question. When we measure the age of the universe, why do we assume that at creation the heavy elements like uranium predominated and not the simplest ones, hydrogen and helium? . . . It is philosophically simpler to assume that all started—if there ever was a start—with the most elementary elements. A catastrophic event or many such events were necessary to build uranium from hydrogen . . ." [418]. Again, ignorance in a field where Velikovsky talks at length and often. Astronomers and physicists do indeed assume that all started with the most elementary element, hydrogen. From it were formed the first stars, in which energy was produced as a result of fusion of hydrogen nuclei into larger ones, chiefly (or at first) helium. When the hydrogen is virtually exhausted, the star changes in type to one in which heavier nuclei are formed from helium and the remaining hydrogen, and then by more complicated reactions. The possible histories of various types of stars have been worked out in some detail, and examples of them at various stages in their development have been found in the skies. Eventually, some "burned-out" stars explode, scattering their accumulation of the various elements formed in them. Our solar system is much younger than the universe, and was formed from the debris of old exploded stars as well as from the ubiquitous hydrogen. The solar system, but not the universe, is dated by means of the time taken for uranium to decompose. Cosmologists *do* assume what Velikovsky says they don't but should. Astronomy seems to be the part of physical science that is closest to Velikovsky's interests. Yet he shows himself to be unaware of contemporary cosmological views, even to the extent of being unclear about the distinction between the problem of the age and origin of the universe and that of the age and origin of the solar system.

Red Herrings and Straw Men

Velikovsky does not hesitate to use the specialists' jargon and the tone of authority, but his statements about the methods, facts, and laws of science cover a range from correct, through almost correct, all the way to quite wrong. As a result the layman is virtually bound to be misled. Consider the following examples.

". . . mutations . . . a process of spontaneous changes in living nature fundamentally different from the process of evolution . . . postulated by Darwin" [431]. Darwin did not concern himself with genetic mutations because knowledge of genetics came much later than Darwin. But modern biologists see mutation as providing the variability (or some of it) upon which Darwin's natural selection acts to produce evolution. Mutation and natural selection go hand in hand in biological science, and it is quite misleading for Velikovsky to write as though mutation and evolution were in some way contradictory. Further, he is wrong in ascribing "evolution" to Darwin—many evolutionists preceded Darwin, whose notable contribution was to provide the idea for a mechanism, that of natural selection, through which the process of evolution could be understood.

"The undulatory theory of light transmitted by waves in the ether. . . . The ether was discarded in the Special Theory of Relativity, and Einstein embraced the quanta theory of light . . ." [441]. This implies that the "undulatory" and "quanta" theories were antithetical. In fact, all scientists (since, and including, Einstein) view light as a phenomenon that is neither particulate ("quanta") nor wavelike ("undulatory") but whose behavior in some situations is mathematically describable by equations for wave motions (the phenomena of refraction and diffraction) and in other situations by equations for particles (when energy is transferred).

"W. H. McCrea . . . came to the conclusion that . . . no planet could have been formed inside the Jovian orbit . . ." [438]. McCrea's argument [243], based on the Roche limit (closest approach of two bodies under gravitational influence without disintegration), is that no planet having a density equal to, or less than, that of Jupiter could have formed inside the Jovian orbit. Planets of greater density could—and all the inner planets in fact do have densities greater by a factor of about four than that of Jupiter (see Chapter 6, *Venus Is Anomalous*). The manner in which Velikovsky quotes McCrea is quite misleading, since it does not make clear that the question of density is crucial. The effect of Velikovsky's reference is to imply to the reader that the existence of planets inside the Jovian orbit is a mystery in the light of conventional theories, whereas it is not.

". . . in the deposits of the Gulf of Mexico the age of oil is measured in thousands of years, not millions. . . . This destroys the main argument the geologists have raised against the theory of exogenous origin of some deposits of oil (*Worlds in Collision* pages 53–58, 369)" [431]. Not so: one of the chief arguments against Velikovsky's suggestion that oil plummeted down and then seeped into deposits is that oil, lighter than water, has in fact been pressed *upward* through porous formations to be held in basins under nonporous rocks, and could not have seeped *down* through the latter [214]. Further, it is misleading to give *Worlds in Collision* as reference for "the" theory of exogenous origin of oil—one hypothesis held by a number of conventional scientists is that hydrocarbons were present in the material from which the planets formed and that some oil deposits might be of such "exogenous" origin, although not those with such recent dates of formation. And here again Velikovsky is misleading in implying that the date of "thousands of years" makes his theory in any way more plausible: the dating [375] was about 9,000 years, whereas Velikovsky's scenario would require a date of about 3,500 years. Beyond that, the method (carbon-14) used for measuring the age is based on the premise that the oil was formed from once-living material (vegetation) under normal terrestrial conditions—a very different history from that proposed by Velikovsky. No date obtained by this method could provide support for Velikovsky's hypothesis unless he first gives a calculation showing the amount of carbon-14 to be expected in oil formed and deposited in the manner he described.

". . . F. Whipple . . . came to the conclusion (1950) that two collisions occurred between these bodies [the asteroids] and a comet, once 4700 years ago and the second time 1500 years ago. . . . These dates . . . are of the same order as those offered in *Worlds in Collision* . . ." [431]. But Whipple's conclusion has nothing whatever to do with Velikovsky's scenario. The dates of Velikovsky's catastrophes are about 3,500 and 2,700 years ago—no doubt "of the same order" but still entirely different.

Quoting from Hawkins, Velikovsky says that with "27,060 alignments in a structure designed as an observatory it is surprising to read that 'stars and planets yielded no detectable correlation.' . . . Stonehenge, if it was used for astronomical observations, must have been put together . . . under a different celestial order" [419]—which would remove the existence of Stonehenge as a stumbling block to Velikovsky's scenario. But Hawkins did find correlations with the sun and the moon, and it was observations of those bodies only that he claimed as a possible purpose for Stonehenge. Those correlations exist if the apparent motions of the sun and moon were the same when Stonehenge was built as now, certainly a stumbling block for Velikovsky's views. By first making light of the accuracy of Hawkins's correlations with sun and moon, and then quoting the admitted *but irrelevant* lack of correlation with the motions of other bodies, Velikovsky misleads as to the thrust of Hawkins's arguments.

Proven

Not only in *Cosmos without Gravitation* but also in *Worlds in Collision* and later writings, Velikovsky displays a lack of understanding of chemistry, physics, and astronomy. Yet he discusses these subjects in a manner that would convey, to a layman, an apparent familiarity with these fields; to a scientist, Velikovsky's ignorance is evident.

Indeed, Velikovsky's whole approach is quite unscientific. He builds a long chain of assumptions without testing the validity of the individual links in the chain (see Chapter 6). He "borrows credence" [408:preface] for his revised chronology—expecting others, apparently, to take on trust a negation of what is accepted by historians in order to follow the author on further unorthodox speculations about astronomy. Velikovsky's reasoning is simplistic in the extreme: as for example his implicit assumptions [127] that "similarity of form reflects simultaneity of occurrence"; that "events are sudden because their effects are large"; that one can infer "world-wide events from local catastrophes." His mode of reasoning is like that of Burnet in the seventeenth century, seeking natural explanations for natural events by taking as literal truth the accounts in ancient records [127, 128]. That approach was a rational one at that time in view of what was then known, but it is no longer so three centuries later, when it becomes necessary to use what science has learned in the meantime. Simplicism also vitiates any significance the advance claims might otherwise have; the hand-waving form of the advance claims demonstrates how out of touch Velikovsky is with actual scientific work. No scientist would think that he had said something significant merely by speculating that Venus is hot, or that Jupiter might emit radio signals, unless he also gave a logical chain of reasoning that led—at least in principle—to something more specific: actual temperatures and frequencies, say.

In *Cosmos without Gravitation* Velikovsky confuses principles with applications of principles, and equates the neglect of a small effect with a statement or implication that the effect, or its cause, is not present at all. He also appears to believe that a theory, or a set of theories, is necessarily so inadequate that it should be discarded simply if there exist some phenomena that have not yet been explained. His use of references and quotations also disturbs me deeply. In a number of instances his usage leaves the implication that a particular reference supports his argument, when in fact the very opposite is the case. Further, he has a penchant for quoting older, rather than recent, work; in scientific activity it is accepted practice to cite also the most recent relevant publications.

Velikovsky, then, is not only ignorant of the facts and theories of modern physical science; his whole approach is not that of a scientist. He does not weigh his evidence—he merely picks and chooses, retaining what he likes and discarding what he does not (as with the books of Humphreys and Jeans); he does not adduce *independent* tests of validity—tests independent of the basic premise. My conclusion inevitably is that Velikovsky's work is not science: it is not amenable to scientific discussion or testing. Since, however, Velikovsky presents his work as scientifically valid, there is warrant to describe him as a pseudo-scientist. He illustrates what Pope was talking about [322]:

> A little learning is a dang'rous thing;
> Drink deep, or taste not the Pierian Spring:
> There shallow Draughts intoxicate the Brain,
> And drinking largely sobers us again.

To my own satisfaction, then, I have concluded that Velikovsky's ideas about matters of natural science are not worth taking seriously; I have set out my reasoning in order to obtain reasoned assent to that view. But it is the fact that other individuals, also knowledgeable about science, do not share my view (see Chapter 5, *Velikovskian Science and Its Practitioners* and *Scholarly Support*). That sort of disagreement occurs frequently on topics on the fringes of science—regarding extrasensory perception, for example. In such disagreements it is common to find the proponents of unorthodox views described as pseudo-scientists, cranks, or crackpots, as some people have labeled Velikovsky. Name calling of that sort is not an appropriate mode of reasoned argument: it does not address the substantive issues. History teaches that ideas once labeled "crackpot" have sometimes turned out to be worthwhile; "cranks" have sometimes been vindicated. My conclusion that Velikovsky is a pseudo-scientist means only that he works outside the currently accepted borders of science; it does not mean that he is wrong in his assertions. In the following chapter I examine the manner in which name calling is used to discredit unorthodox views, particularly how this occurred in the Velikovsky affair.

8

Pseudo-Scientists, Cranks, Crackpots

We seldom attribute common sense except to those who agree with us.

—de la Rochefoucauld

According to the *Oxford English Dictionary,* a crank is "*(U.S. colloquialism)* a person with a mental twist; one who is apt to take up notions or impracticable projects; especially one who is enthusiastically possessed by a particular crotchet or hobby; an eccentric, a monomaniac." It is common to so label people who express whole-hearted, thoroughgoing belief in: astrology; fortune telling by the reading of palms, cards, tea leaves; numerology and pyramidology; extrasensory perception; the reality of flying saucers, especially those carrying intelligent visitors from elsewhere than earth; reincarnation; spiritualism; the existence of sea serpents and Loch Ness monsters. The list could be expanded greatly; many books and articles have dealt with relatively well-known cranks and the ideas they espoused.

Martin Gardner [113] discusses "the curious theories of modern pseudo-scientists and the strange, amusing and alarming cults that surround them—a study in human gullibility." We are regaled with the ludicrous ideas that the earth is flat or that it is hollow, with us on the inside; with Hans Hoerbiger's theory of successive moons and the prevalence of ice in space; with tales of catastrophic encounters between the earth and a comet (Whiston, Donnelly, Velikovsky); with antigravity research, dowsing, antirelativity theorists, medical quacks, food faddists, and so on. Daniel Cohen recounts *Myths of the Space Age* [51]: astrology, extrasensory perception, reincarnation, flying saucers, pre-Columbian discoverers of America, Immanuel Velikovsky, Loch Ness monsters, yetis. Christopher Evans [104] explores modern religious cults, in particular the dianetics and scientology of L. Ron Hubbard. John Sladek surveys *The New Apocrypha* [373]: if science says one thing, the new Apocrypha state the reverse—thus Atlantis, Bacon's ciphers in Shakespeare's plays, homeopathy, Loch Ness, Nostradamus, perpetual motion, Velikovsky, Wilhelm Reich, Zen macrobiotics. Ronald Story [388] and Peter White [462] have exploded the fanciful and fancied "evidence" that early civilizations were influenced by intelligent extraterrestrial beings. Charles Fair [106] tells of the fallacies of ESP, UFOs, and Velikovsky. An early classic in this field is *Foibles and Fallacies of Science* [147], which ranges widely over astrology, perpetual motion, divination, attempts to refute the Newtonian theory of gravity. Recent years have brought an increasing number of books in this genre, as well as journals and societies [372, 376, 474] concerned with anomalies and unorthodoxies.

Patrick Moore [256] covers cranky notions regarding cosmology in particular (flat, hollow, and other peculiar models of the earth; Hoerbiger, Velikovsky, von Daeniken, UFOs, astrology, ESP, Atlantis). Moore prefers to speak not of "cranks" but of "Independent Thinkers":

> . . . The difference is important. The Independent Thinker is a genuine, well-meaning person, who is not hidebound by convention, and who is always ready to strike out on a line of his own—frequently, though not always, in the face of all the evidence. He is ready to face ridicule; he believes himself to be in the right, and he cannot be deterred. In some respects he is a rather special kind of person, though generally speaking he is conventional enough except in his one particular line of thought. He may or may not be scientifically qualified. . . . All share the wish to inquire, and—this is the vital fact—all are anxious to do something really useful. . . .

> In my personal investigations into the world of Independent Thought, I have been impressed with the charm, the courtesy and the patience of those concerned. They are so totally unlike the religious cranks, the money-grubbers, and the more extreme child psychiatrists whose doctrines have caused so much havoc during the past couple of decades. . . . I am a conventionalist, and also somewhat sceptical, so that in a way I am acting as devil's advocate. Now and then the Independent Thinker may be right when others are wrong. More often he is grasping hold of a totally false idea. . . . [256:14–15]

Moore's distinction between Independent Thinker and crank is based not on the nature of the ideas involved but on the personalities of those involved—both espouse ideas that are more or less foolish, but Independent Thinkers are rather lovable whereas cranks are not. I would plead for a distinction along somewhat different lines: let us by all means criticize as foolish those notions that seem to us foolish but, no matter whether we like a particular individual or not, let us distinguish between the person and the ideas. Foolish ideas do not make a fool—if they did, we could all rightly be called fools.

The books just mentioned are representative of a burgeoning literature [3, 32, 47, 69, 70, 114, 237, 329] that aims to debunk unorthodox ideas. Typically, cranky notions are described but the supposed errors in them are rarely demonstrated—the authors assume that the preponderance of their readers will concur, without requiring proofs, that the notions are indeed cranky. The effect is that we are presented rather dogmatically with descriptions of purported absurdities, with the clear implication that, if we have any sense, we also will view them as absurdities. Thus Sprague de Camp [67] says that "taking apart Velikovsky, the pyramidologists, and the hollow-earthians is like shooting fish in a bathtub. . . ."

I imagine myself in a typically small bathroom, and wonder whether shooting at those fish is really such an easy and safe avocation. Might not the stray bullet ricochet with disastrous results?

These writers and many others—as well as most of the rest of us—have few qualms about their (our) certainty that Velikovsky, among others, is a crank. How can one be so sure, when others—equally respectable in accomplishments and intellect—are not sure? Velikovsky has been taken seriously by Alfred de Grazia, a professor of political science with a respectable reputation in his field; by C. J. Ransom, a physicist; by Ralph Juergens, an engineer. He was befriended by Harry Hess, one of America's foremost geologists; by literary critics, humanists, social scientists. Who is qualified to make the judgment? To say, not "I think Velikovsky's ideas are not valid" but, with the implication of objectivity, "Velikovsky is a crank"?

In fact, of course, no one is qualified to make such a sweeping *ex cathedra* pronouncement about Velikovsky or about a number of the other subjects listed above. We see in action here the tendency for human beings to speak with the language of certitude when all that is warranted is the expression of a personal judgment of plausibility or probability. Martin Gardner, in the second edition of his book, describes [113:preface] the letters he received about the first edition: "most of these correspondents objected to one chapter only, thinking all the others excellent," and, of course, it was not always the same chapter to which the various correspondents objected.

I would myself agree with the judgment that most of the "crank" ideas mentioned in the foregoing are indeed implausible in the extreme. But I found it no easy matter to reach an opinion about Velikovsky. I hold one now not because Gardner, de Camp, Asimov, or anyone else said so but because of my own examination of most of the material about Velikovsky and by him. The critics of Velikovsky did little, if anything, to help me reach my conclusions—their *assertions* were of no assistance to me in my search for *evidence* and *proof*.

Further, when an author lumps together Velikovsky and the Loch Ness monster (see Cohen [51] and Sladek [373], above; Nichols [274], in Chapter 1, *The Battle Begins;* Goudsmit [126] and Spencer Jones [164], below), that author's credibility is lowered in my estimation. Just as from personal inquiry I believe Velikovsky to be a pseudo-scientist, so also on the basis of my personal inquiries I believe in the existence of Loch Ness monsters ("Nessies") and of sea serpents. In Loch Ness there is a breeding population of large aquatic animals with powerful flippers, long thin necks, and bulky humped bodies, animals not as yet known to "science." I believe that because I myself have examined the evidence of eyewitnesses, of photographs, of sonar observations.[1] So, someone who lumps my Nessies together with the case of Velikovsky loses credibility in my eyes; it indicates to me that he probably takes his opinions at second hand or after only cursory reading. And when someone classes Velikovsky (or anybody else) as a crank, I expect that writer or critic to have his facts straight, to have reached an informed judgment that he is passing on to me. Lamentably, we are all fallible, and those who label Velikovsky "crank" are as fallible as the rest of us—even if they may seem to be unaware of that. Dogmatic labeling of people as cranks and of ideas as cranky abounds.

Menzel [247] thought that the near-unanimity of scientists proved Velikovsky to be a crank. Margolis [236] was clear that Velikovsky's concepts are "plain hokum." The physicist Gell-Mann [116] classed Velikovsky with astrology and palmistry. Another physicist spoke of "public interest in the abstract, the occult, in extrasensory experiences and the Loch Ness monster . . . the lunatic fringe includes some physicists" [127]. Sir Harold Spencer Jones [164] was confident that others suffered from hallucinations: "not infrequently . . . when there is a report of something having been seen which is mysterious and outside ordinary experi-

1. The most recent, reliable, and comprehensive account is by Witchell [467]. Good bibliographies are given by Costello [59] and Mackal [229]. The underwater photographs and some sonar results from 1972 and 1975 have been published by Rines, Wyckoff, Edgerton, and Klein [339]. The only extant moving film of one of these creatures was taken in 1960 by Tim Dinsdale, who has written several books [86–89] about the search and the findings over the last two decades.

ence, other people begin to think that they see the same thing . . . reports of the Loch Ness Monster provide an instance. In the case of the flying saucers, something similar seems to have occurred. . . ." We have been told that "it is surprisingly easy to be certain in the majority of cases that one is not withholding recognition from a second Einstein" [340]; that we can "conclude with an unusually high degree of safety that Velikovsky's theories are pseudoscientific nonsense . . ." [38]. An astronomer assured us of his infallibility in this respect: "the library . . . has a shelf . . . reserved for the screwball fringe of science. . . . Through a period of nearly thirty years I have consigned to this shelf a large number of books. It has never yet become necessary for me to remove one and assign to it a more honorable place . . ." [393]. But all this confident labeling does not help those of us who want to make up our minds on the basis of evidence rather than on the basis of someone else's opinion. It is of little use to read that Velikovsky is like the cranks Donnelly, Voliva, Hoerbiger, Hubbard, etc. [68], when one can also read that "a thin, but diffuse line . . . distinguishes Velikovsky from . . . Donnelly . . . Horbiger" and others [166], and that "the implication that Velikovsky's scholarship is comparable to the writings of men such as Voliva and Hubbard is ludicrous" [10].

We are familiar—in principle at least—with the fact that one man's sense is another's nonsense. If we just accept the opinion of one who labels cranks, we may as easily be nonsensible as sensible. Boring [34] reminds us that it is very difficult to get the consensual agreement of nearly all wise men and that, in any case, wisdom in one field does not usually transfer to another. It is much easier to label a man a crank than to prove that he is one or that his pet notion is cranky (as George Bernard Shaw illustrated with his accustomed flair for the case of a flat-earther [367:360–61]). Moreover, one finds that the most confident of crank labelers are capable of being wrong, of making statements that are misleading or untrue or that show ignorance of the subject being discussed.

Sprague de Camp [67] said that "when Velikovsky

quotes Herodotus . . . and Hesiod . . . and Isaiah . . . you have only to turn to the books cited to learn that Herodotus and Hesiod and Isaiah said nothing of the sort." Well, it may be that Velikovosky's interpretation of what those authors said is far-fetched, perhaps even clearly invalid, but it is absurd to state categorically that they said "nothing of the sort" (see Chapter 4, *Summing Up the Second Engagement*). De Camp's statement merely indicates to me that de Camp has not himself, open-mindedly, judged Velikovsky's writings and the arguments about those references in particular. This issue cannot be settled quite so easily, and de Camp's attitude does not commend him to me as one whose judgment is to be relied upon. Martin Gardner [113:32–35] talked of "Professor Stewart's devastating criticism" of Velikovsky—ignoring the published evidence [61, 470] that many people do not judge that criticism to be at all devastating. Gardner stated that "*Worlds in Collision* is no longer taken seriously," but it is taken just as seriously now as it ever was by those who were not of Gardner's mind in the first place. In these statements Gardner is indulging in wishful thinking, not accurately portraying a situation.

As a final example, Asimov [17] wrote that "Macmillan planned to publish the book as part of its *textbook* line . . . as a scientific textbook." That is quite wrong (see Latham's account [213:71–77]), on a significant point. Asimov also said that "Velikovsky doesn't accept the laws of motion, the law of conservation of angular momentum, the law of conservation of energy and other such trivialities." That is an irresponsible statement: Velikovsky has never expressed a disbelief in those laws—he has argued that certain events that he postulates would not contravene those laws, whereas others have said (but not proved) that those events would contravene those laws. Velikovsky *has* expressed skepticism about the theory of gravitation. I could agree if Asimov were to point to evidence that Velikovsky does not grasp the implications of some fundamental points of physical science (see Chapter 7). As it is, though, Asimov's account is so inaccurate and misleading that he stands revealed of either ignorance or misrepresentation when he writes about the Velikovsky affair.

Label with Care

The important point is that it is not easy to be certain—*objectively* speaking—whether or not a particular individual is a crank. I do believe that Velikovsky is a crank in the matter of gravitation, but I also believe that others have labeled him a crank in the matter without sufficient evidence, without a properly informed view of the whole affair. I do believe that Loch Ness monsters are real animals, and others have labeled me a crank for that reason—without having seen all the evidence that I have seen. The fact is that one can very rarely be quite certain that use of the label "crank" is justified. And when those who use the term are pressed, they will generally appear to concede that point.

Asimov [17] admits that "almost everyone has a touch of the CP [crack-pot] in him—including most certainly myself," but he then tries to back away from that concession to maintain that nevertheless *he* knows a real crank when he sees one: "There are . . . certain distinguishing marks by which you can tell a *far-gone* CP; one in whom CP-ness tends to drown out everything else. And when I use the term in this article, it is the far-gone variety to which I refer."

Sladek [373:15] makes "the effort . . to distinguish between ideas which are off the beaten track and those which are simply off the rails," but he does not tell us how that effort is, or can be, made. He does give an illustration, however: "The newspaper that carries moonshot pictures [science] on the front page also carries the daily horoscope [pseudo-science] inside. Heart transplants [science] and faith-healing [pseudo] are treated with equal seriousness by all but the most responsible papers—and even these can seldom resist a hot Loch Ness [pseudo] story" [373:13]. Thus Sladek's criteria or methods have led him into error concerning Loch Ness. He also castigates belief in faith healing—would he deny the reality of psychosomatic illnesses and cures thereof?—yet any distinction between psychosomatic ills and faith healing is surely more one of semantics than of substance.

Gardner [113:50] "can grant that Einstein may be wrong, and . . . a faint (*very* faint) possibility that Velikovsky may be right," but he immediately proceeds to translate that estimate of *probability* into a statement that is intended to convey conviction, *certitude:* "The extremes of the continuum are so great that we are justified in labeling one a scientist and the other a pseudo-scientist." Here Gardner is also mistaken on a central issue. One can indeed legitimately call Velikovsky a pseudo-scientist in the sense that he is not a scientist and his work does not fall within the scope of what we mean by scientific activity, but that does not necessarily mean that Velikovsky is not right.

Sprague de Camp [67] is clear in principle about the difficulties involved: "Picking winning scientific theories is not unlike picking winning horses. Some do it better than others, but how much of that is luck and how much is brains is known to nobody"; and "there is *no* method of judging scientific theories that is *both* easy and sure, and no method at all that is completely certain. . . ." But, in practice, de Camp can arrive at conviction after all: "even if no scientific theory is ever *completely* 'established,' it often becomes established quite well enough for all ordinary human purposes"; "it is conceivable that tomorrow somebody will really prove that the earth is flat, or that werewolves exist, or that diseases are caused, not by germs, but by astrological influences. But I don't think the chance is big enough to worry about." De Camp may be right that theories can be established well enough for ordinary purposes—just so long as he realizes that those purposes are to use the theory for the sorts of cases in which it has worked satisfactorily in the past and thus will very likely continue to in the future. He is wrong if he intends thereby to convey (as he indeed implicitly conveys) that the theory is therefore right, true. That a theory is usable does not make it true. Moreover, de Camp's use of examples is disingenuous—he cites issues on which there is a broad consensus that the theories or ideas are false; one cannot transpose that analogy to a situation in which there is no general consensus.

When pressed, then, those who write about cranks will admit their own fallibility, or at least the general possibility of fallibility, in this labeling. But the admission is usually made in terms, or with qualifications, that are intended to convey that this fallibility really does not apply in the particular case that is being discussed. In effect, we are told, "I know that I might not always be right—but I do happen to be right in this instance."

An attempt to define a procedure for identifying cranks was made by Lafleur [207]. Characteristically, according to Lafleur, the crank

1. Lacks awareness or understanding of the theories that he opposes.

2. Contradicts existing theory by maintaining the correctness of a hypothesis without showing that the latter has more *raison d'être* than the former.

3. Opposes established theory not only in the particular discipline in which his idea falls but also in a range of disciplines.

4. Proposes a theory that is not a satisfactory substitute for the old.

5. Holds opinions that do not lead to a satisfactory overall world-view.

6. Fails, compared to accepted authority, in the prediction on the basis of his theory of verifiable consequences, particularly where such consequences are capable of being quantitatively (mathematically) expressed.

7. Tends to accept minority views in a variety of areas and to overemphasize the fallibility of science.

On the face of it, we have here a set of workable criteria;[2] most of us would be agreeable to labeling as crank someone who meets all these criteria. That labeling would not neces-

2. In a posthumously published book [416], Velikovsky destroys in witty fashion Lafleur's attempt to distinguish genuine revolutionary theories from cranky ones.

sarily be therefore correct, however. A man who is right about something, but prematurely so—before the discipline is ready for the new idea, theory, or phenomenon—would also seem to fulfill several of these criteria. Such premature discoveries, like the works of "true" cranks, tend to be ignored by the majority of working professionals (see Chapter 15, *Premature Discoveries*). Moreover, even if these seem to be good criteria in themselves, their application to any particular case involves making judgments rather than looking at objective facts. Lafleur himself, for example, maintains that Velikovsky qualifies as a crank under most if not all of the seven criteria. Yet supporters of Velikovsky continue to make much of the success of his predictions (criterion 6), of the interdisciplinary character of his work (number 5), and of his grasp of the content of many disciplines (number 1), and they do not agree that his theories are incompatible with valid current views.

This difficulty of applying generalizations in specific cases makes even the most plausible set of principles of academic interest only. Bernstein [29] noted the following: the crank manuscript solves everything; the ideas are not in any way connected to conventional knowledge, are (in Pauli's phrase) "not even wrong"; cranks are humorless, determined to bring the newspapers in, afraid that everyone is ready to steal their ideas, not really interested in the rough-and-tumble of criticism that goes with doing science. All these points have a ring of authenticity, but in any given instance one would have difficulties: the manuscript solves a lot but not really everything; the "crank" and his supporters will certainly find *some* connection to conventional knowledge; who defines humor? and so on.

An interesting discussion of how to distinguish science from pseudo-science was given by Langmuir [208]. He dealt with some cases of "discoveries" in the mainstream of science (or by workers who were in the mainstream), cases in which the discoveries later turned out to be quite unfounded, spurious. Langmuir spoke of "pathological science" in this connection, and listed the symptoms:

1. The maximum effect that is observed is produced by a causative agent of barely detectable intensity, and the magnitude of the effect is substantially independent of the intensity of the cause.

2. The effect is of a magnitude that remains close to the limit of detectability; or, many measurements are necessary because of the very low statistical significance of the results.

3. Claims of great accuracy.

4. Fantastic theories contrary to experience.

5. Criticisms are met by *ad hoc* excuses thought up on the spur of the moment.

6. Ratio of supporters to critics rises up to somewhere near 50% and then falls gradually to oblivion.

Commentators pointed out that some genuine discoveries would have met these criteria when those discoveries were first made—the test of time is therefore an important one. According to Langmuir, in the case of genuine discoveries the effects observed are quickly found to be measurable well above "threshold" under suitable conditions, and the reproducibility of the phenomenon improves accordingly. In pathological science the phenomena do not become more reproducible as time goes by—in Rhine's series of card guesses to establish ESP, for example, or with reports of UFOs. The test of time may, however, need to be long, indeed: the Loch Ness monsters have been sought more or less systematically at various times since the 1930s, with no improvement in reproducibility. Only with the advent of sonar and of underwater photography has the rate at which evidence is being obtained shown any signs of increasing, and even then not very markedly.

The difficulties under discussion were recognized by Gruenberger [136], who proposed that one rank individuals according to their scores on a series of criteria that are given varying weights and that attempt to describe how a scientist ideally might behave. These criteria include that the work be publicly verifiable (repeatable, at least in principle, by others); success in predicting; controlled experimentation (admittedly not applicable to such observational sciences as astronomy); the use of Occam's razor (using the simplest possible hypothesis rather than a more complicated one); fruitfulness; and sanction by established authorities in the discipline. Criteria pointing in the opposite direction include use of the Fulton *non sequitur* (they laughed at Fulton, but he was right; they laugh at me—*therefore* I am right);[3] paranoia; inflated opinion of the importance of the work; overindulgence in the use of statistics. Gruenberger applied his scale to some hypothetical instances: a universally recognized physicist scored 97 (out of 100), a researcher in ESP scored 38, and a dowser scored 28.

Although Gruenberger's discussion is interesting, his scale does not lead to the certain identification of cranks. First, all that the scale measures is how an individual fits into the accepted norms of scientific activity—not, what really interests us, whether or not the ideas may be valid. Second, the assigned scores for each item are wholly subjective. Third, there is no overall score above which one can place "scientists" and below which we have "cranks." Indeed, Gruenberger did not propose that this metric be used to arrive at an infallible judgment, and his discussion underlines the fact that one is dealing not with two classes of people—scientists and cranks—but with human beings, whose attributes on any such metric will fall on a continuum.

We are left, then, with the conclusion that we can judge only probabilities. Thereafter each one of us makes a subjective judgment, converting probability into subjective certainty—deciding "scientist," or "crank," or perhaps (and quite rarely) "not proven." This is no easy lesson to digest. It is perhaps not difficult to admit that others may make errors of judgment, but what we must face is that we, ourselves, may do so even when we are most certain that we are right.

Sprague de Camp [67] reminds us that Galileo, rejected by the authorities but later shown to be right, was also a crank on the matter of planetary orbits—he insisted on the

3. The laughter clearly does not imply that one is right, but I. J. Good [120] points out that if an idea is not laughed at, then it is probably not very original.

principle of circularity and rejected Kepler's notion of elliptical orbits, which we now accept. Francis Bacon, one of the respected founders of modern science, was scathing about the heliocentric idea. C. P. Smyth, an acknowledged pioneer in spectroscopic studies, was a crank in the matter of the Pyramids. The names of those who are now respected as great innovators, but who were labeled by their contemporaries as cranks, charlatans, or wicked, are legion [113, 125]: Horace Wells (anesthesia with nitrous oxide), Semmelweiss (sterilized environment for the delivery of babies), Pasteur, Roentgen, Lister, and so on. Moreover, Gardner [113:8–9] reminds us that even charlatans may sometimes be right: Mesmer was something of a crank and a charlatan, but hypnosis—Mesmerism—is a real phenomenon; Robert Mayer, a psychotic, discovered the law of conservation of energy.

Cranks and Geniuses

> The reasonable man adapts himself to the world; the unreasonable one persists in trying to adapt the world to himself. Therefore all progress depends on the unreasonable man.
>
> —George Bernard Shaw

> . . . Out of egoism are derived the drive and enthusiasm that lead men to undertake research, to keep at it, to publish the results, to keep promoting the knowledge and use of these results. Also out of egoism is derived the emotional support for the cranks, for Velikovsky's collision of two worlds, for Hubbard's dyanetics, for every scientist who still holds on to a theory after the weight of evidence no longer justifies it. [33]

> Like Galileo, Velikovsky brought much of the opposition on his own head by failing to understand the rules of the game and by arrogance which developed into paranoia as every turn he made was blocked by the conventionalists. [235]

> I don't know the extent to which Velikovsky is genuine or spurious. If he were altogether the one without the other, he

would be unlike most remarkable pioneers. After all, Columbus was a bit of a promoter and adventurer. . . . [64]

There is rarely a certain way of identifying a crank. The cited quotations indicate why, quite apart from the difficulty of analyzing the substantive issues: psychological attributes of drive, persistence, egotism, subjective certainty (what Boring [34] has so aptly described as "intractable enthusiasm") are of necessity present in men who try to force their unorthodox views on the unreceptive orthodox ones. Such iconoclasts are geniuses or they are cranks. Contemporaneously, one can rarely distinguish between them on the basis of the substantive issues, and one can certainly not distinguish them on the basis of personal characteristics.[4]

That, however, ought not to lead us into the Fulton *non sequitur,* into being ready to rank a man as genius merely because he is heterodox and persistent: "We know that many fundamental beliefs of modern science arose as heretical speculations advanced by nonprofessionals. Yet history provides a biased filter for our judgment. We sing praises to the unorthodox hero, but for each successful heretic, there are a hundred forgotten men . . ." [127]. We have all heard of the great heretics who proved to be right, but we simply never

4. There are, of course, extreme cases. Cohen [51: 2–3] has given the following description of cranklike attributes, which has the ring of truth to it: ". . . a head-to-head collision with a confirmed crank can be a really frightful experience. Suddenly one must deal with a mind that cares little for evidence and even less for logic. The crank seems to have twice as many hours in the day as an ordinary person does to gather information, usually obscure and almost always irrelevant, to support his obsessive beliefs. In any argument, he flings this information at his opponents in great handfuls. No sooner has the critic knocked down one set of propositions than another set comes flying at him and that, too, has to be dealt with. The crank can produce a seemingly endless stream of books, articles, and letters, and most of all he can talk, talk, talk. A confrontation like this can be agonizingly frustrating and unbelievably exhausting. Bertrand Russell once observed that the only way to deal successfully with a true crank is to counter his preposterous assertions with even more preposterous ones, until he is driven away, thinking that you are the crank. Few, however, have that much energy or imagination." In practice, however, one does not usually come across such clear-cut examples.

hear about the forgotten ones who were wrong, even though they have been so much more numerous. There is admittedly a small literature dealing with the more colorful cranks, but then there still remain all the people who—as it turned out later, incorrectly—pushed unorthodox views inside the disciplines themselves; those are truly forgotten men. Even the literature dealing with cranks is not widely known: that is presumably the reason why so few recognized the spuriousness of the claim of extraordinary originality made for Velikovsky's ideas—who had heard of Whiston and Donnelly?

Just as we usually cannot (except by hindsight) distinguish the crank from the genius on the basis of his work or of his personal characteristics, so we also cannot make the distinction on the basis of the natures and actions of the followers that he attracts. It is evident that one needed to tread a narrow path to remain an acknowledged associate of Velikovsky: he "demanded" certain changes in the structure of the discussion groups of his work [421]; he virtually ensured the demise of *Pensée* by refusing to contribute further articles to it because of disagreement with its editorial policy [448] (even though that policy was anything but anti-Velikovsky); "as even his friends will admit, [he] is not easy to get along with . . ." [166]. But a similar attitude has been and is shown by many who are acknowledged leaders of an orthodox persuasion. There are not a few leading scientists who demand from their associates absolute adherence to their viewpoint and who simply sever connections with those who come to hold modified, let alone different, opinions. Freud, who has had some special interest for Velikovsky, demanded the strictest personal and ideological loyalty from his associates.

Also, one cannot legitimately put at the leader's doorstep all the miscarriages, ineptitudes, errors, and the like perpetrated by his followers—most particularly not those perpetrated by individuals who blaze their own path while professing still to be adherents, who bowdlerize and cannibalize the leader's views. Thus, that disciples of Edgar Cayce have, in a sense, "adopted" Velikovsky [234] is no doing of Velikovsky's, and it would be quite wrong to judge Velikovsky's ideas on the basis of such an association. Many great innovators have lived to see their views bowdlerized, misinterpreted, and abused. One cannot hold Einstein responsible for all the metaphysical and pseudo-philosophical speculations ostensibly based on "relativity." Neither can one lay at Heisenberg's door all the naive Sunday sermons in which his uncertainty principle was given as scientific proof of the reality of free will in human beings. "By their fruits ye shall know them": perhaps and eventually, but certainly not by their hangers-on.

The Moral of the Story

In such complicated matters as the Velikovsky affair, objective certainty is simply not to be attained, at least not quickly—history's verdict may approach it. There is "no formula for distinguishing the potential contributor from the pretentious crackpot. At best [one] . . . can distinguish those who are working outside the bounds of normal science . . ." [125:251]. When the accepted, conventional scientific wisdom embodied in contemporary experts pronounces a man a crank, that judgment may be wrong. Conversely, however, the mere fact that an individual is labeled a crank does not even begin to suggest that he might be one of those very few who later turn out to have been right after all. The statements of those who criticize the crank are not necessarily correct, even if their judgment later turns out to have been correct. I have shown that Velikovsky is ignorant of physical science to a pronounced extent, that his criticisms of the theory of gravitation are without substance; in the matter of gravity Velikovsky is quite wrong. But much of the writing that purported to show him to be a crank contains errors, logical *non sequiturs,* wrong and misleading statements; the case made against Velikovsky by his critics was neither sound nor comprehensive.

In the foregoing I have shown that Velikovsky's ideas about physics, chemistry, and astronomy are in large measure invalid and uninformed. But even then I cannot say that Velikovsky's cosmic scenario is certainly wrong; I can only say, having given my justification for saying it, that I believe that it is highly unlikely that Velikovsky's scenario could be correct. Still less can I state that his revisions of chronology are invalid—that case has not yet been made by anyone. He is, I believe, a pseudo-scientist, but he is not necessarily wrong. I call Velikovsky a pseudo-scientist because that is a commonly used term. It has surely become clear in the preceding discussion how careful one had better be in using that term, defining and qualifying it as appropriate to the particular case. Pseudo-scientist, crackpot, crank—these are pejorative terms. If we can show an idea to be wrong, why not leave it at that? Why insult the man who put forward the idea?

Perhaps a part of the reason is that, as we have seen, it is not easy to prove the case. By using the pejorative term, we add the force of emotional conviction, to compensate for the lack of logical completeness of our case. Being unable to achieve agreement through discussion—since Velikovsky and his supporters, and others on the sidelines, have never agreed that he has been disproved, not even on relatively minor and peripheral points—we use the big emotional guns of name calling.

I should like at this juncture to disclaim any intention to insult or derogate Velikovsky the man. Where I have a quarrel with him, I have tried to make it explicit and limited to statements on specific issues—for instance, that he is ignorant of physical science but does not recognize that he is, and seeks to speak authoritatively in that field. I used the term "crank" in part to give focus to the preceding discussions; where I use it subsequently, it will be intended strictly in the sense of one who is "enthusiastically possessed by a particular crotchet or hobby," and with the proviso that I myself do not take umbrage when people call me a crank about the Loch Ness monster, for example.

In the next chapter I shall outline reasons why I disbelieve Velikovsky and reject his ideas, reasons quite apart from his lack of competence in the physical sciences. I shall endeavor to show that various aspects of the affair reveal Velikovsky to have characteristics that might well make one hesitant to take his work seriously, even without a detailed examination of the technical issues involved.

9

A Nontechnical Case against Velikovsky

> Commonly we say a judgment falls on a man for something in him we cannot abide.
>
> —John Selden

Self-Importance

Velikovsky projected an overwhelming sense of self-importance, an intellectual arrogance that I find somewhat distasteful: "If these two men of science [Newton, Darwin] are sacrosanct, this book is a heresy . . ." [408:preface]. Thus Velikovsky, at the beginning of *Worlds in Collision,* reveals that he sees himself as comfortable in the company of the greatest men of science that mankind honors. He presents himself as of unquestionable significance: a major heretic.

Another man might have written, say: "Some things in this work seem to contradict ideas that stem from Newton and Darwin. However, I present evidence that appears decisive, and believe that some of those earlier ideas need to be revised." Anyone could write like that. But Velikovsky is not just anyone—he sees himself as one of the truly great: ". . . I was . . . carrying my heresy into a most sacred field, the holy of holies of science, to celestial mechanics . . ." [445]; "no other book in the history of science was made the butt for such abuse and distortion as was *Worlds in Collision*"; those attacks constituted "the most ignominious spectacle and the most deplorable chapter in the history of American letters and of world science . . ." [432].

Linking himself squarely with the great originators, Velikovsky says: "At first a new idea is regarded as not true, and later, when accepted, as not being new" [410:foreword]; "it is not without precedent that most authoritative voices in science should discourage the trail blazer—think of Lord Kelvin . . . who rejected Clerk Maxwell's electromagnetic theory, demeaned Guglielmo Marconi's radiotelegraphy, and . . . proclaimed Wilhelm Konrad Roentgen a charlatan" [445]; "mathematics professors . . . told a large audience 'that . . . Einstein's theory was the greatest hoax in the history of science.' . . . I was honored by a similar lecture at the Hayden Planetarium . . . on the 'greatest hoax in the history of science'" [430].

For Velikovsky, his own work was always the most important point at issue. Thus he referred to the "American Philosophical Society . . . symposium on 'Some Unorthodoxies of Modern Science,' my unorthodoxy being the chief subject on the agenda . . ." [431]. In fact, there was just one paper dealing with Velikovsky on the agenda [290]; another dealt with dowsing [338] and a third with extrasensory perception [187]. In the general introductory paper, entitled "Orthodoxy and Scientific Progress," in more than seven pages one finds only this brief reference to Velikovsky: "Velikovsky's theories are admittedly unorthodox, . . . but their utter rejection is not based on their unorthodoxy, but only on the palpable fact that they are unsupported by a body of reliable data such as is demanded of every new conceptual scheme . . ." [52]. In the summarizing paper [34] of the symposium Velikovsky is given two paragraphs and several other briefer references out of some five pages of text. Yet to Velikovsky, his unorthodoxy was the "chief" subject on the agenda.

Again: Hawkins [144, 145] has proposed that Stonehenge may have been used as an astronomical computer to predict lunar eclipses. Stonehenge, of course, has fascinated many people over the centuries, its purpose remains a matter of opinion, and Hawkins's suggestion has been much debated. The building of Stonehenge predates the catastrophes postulated by Velikovsky, who wrote, "If the ancient alignments are still valid, how could my reconstruction of past events of catastrophic nature . . . be true? Not a small share of the public interest in Hawkin's theory can be attributed to this predicament" [419]. I, for one, have followed with the greatest interest the debate over Hawkins's theory, from long before the time that I had ever heard about Velikovsky or his "predicament," and those who argued with Hawkins never mentioned Velikovsky in this connection. Yet, to Velikovsky, his own presumed role in the matter was of great public significance.

Velikovsky indeed appeared to think that his views were ever-present in the minds of scientists: "I may have even caused retardation in the development of science by making some opponents cling to their unacceptable views because such views may contradict Velikovsky . . ." [445]. He saw criticism of his work as the mark of a conspiracy: "efforts have mounted in the press to deprecate my work . . . and thus to protect the young generation from its influence"; "a stepped-up campaign of deprecation, and often the job is committed to popular writers on scholarly issues, an uncritical avant-garde for those in science who feel they are losing the battle" [432]; "a group of scientists . . . drove many members of academic faculties into clandestine reading of *Worlds in Collision* and correspondence with its author . . ." [410:foreword]. To Velikovsky, progress in science since 1950 involved chiefly the incorporation of Velikovskian views: "the views expounded in *Worlds in Collision* were appropriated piecemeal by those who first opposed them" [410:foreword]; "the astronomers are on the defensive. . . . They asked *me* to participate in the AAAS meeting. I did not ask" [117]; "my work today is no longer heretical. Most of it is incorporated in textbooks and it does not matter whether credit is properly assigned" [445]; "the physicists and astronomers have second thoughts on the validity of my work" [442].

Because of the great accretion of fact, interpretation,

synthesis, methodologies, over the range of intellectual disciplines, it is generally thought that the days are long past when a single individual could be expert in many different fields. So even physicists, for example, recognize that an experimental physicist and a theoretical physicist differ in their degrees of competence to discuss various areas of physics [85:56, 96–98]. But then there is Velikovsky: "I was in the twentieth century a student of the kind there was 100 years ago. But I didn't stop studying up until today . . . I opened and closed the library at Columbia for eight or nine years (certainly I was the greatest exploiter of that institution)" [35].

Within the ranks of those who take Velikovsky's work seriously, Velikovsky himself must of course be the arbiter: "The story of the first four or five years of Cosmos and Chronos [campus discussion groups of Velikovsky's work] and what changes in the structure of the organization I had to demand is a story by itself" [421]—not changes suggested, requested, or put to a vote but changes "I had to demand."

Take My Word for It

In the scholarly world—as is well known, for instance, to assistant professors who do not achieve tenure—credit is rarely given for unpublished work. Only after the work has been favorably passed on by referees and reviewers is there thought to be a *prima facie* case that it might have merit; even then, one awaits the opinions of other experts in the field. Unpublished evidence is not taken as support for any statement or argument. Yet Velikovsky habitually writes as though others should give credence to his unpublished materials. "In a history of earlier catastrophes I shall show . . . [the Pyramids] were not tombs, but royal shelters from natural upheavals . . ." [430]; "the Universal Deluge took place . . . between five and ten thousand years ago, probably closer to the second figure. . . . It is quite possible that the volume of water was more than doubled on earth in this one cataclysm" [438]. As to his reasons for expecting radio signals from

Jupiter, Velikovsky refers us [433] to his unavailable correspondence with Einstein. In *Cosmos without Gravitation* he says, "The theory . . . given here in synopsis is written also in a comprehensive form (1941–43). I arrived at this concept early in 1941 as the result of my research in the history of cosmic upheavals. . . . A number of facts proved to me that . . . gravitational attraction or the weight of objects has changed during human history . . ." [407:21]. I have not been able to find any later reference to this remarkable contention, let alone any evidence for it. And again: "although I have not yet discussed the origin of Venus from Jupiter in any detail, I have already revealed where the escape energy came from [the collision of major planets] . . ." [434]; yet that collision itself has only been asserted, not established—the data on which the statement is predicated have not been published.

Throughout his writings Velikovsky dogmatically asserts the correctness of interpretations that are at best possible.[1] Here are some examples from *Ages in Chaos*:

On p. 45 the Ermitage Papyrus is said to relate "the same story . . . that we now know from the Papyrus Ipuwer, but in a different way . . . as things that are to come. Obviously this indicates only a preference for the literary form of foretelling." The papyrus tells of a perished land, a veiled sun, a dry river, and Velikovsky assures us that "we can recognize" this "description of the changes in nature . . . as belonging to the period when the Israelites roamed in the desert, under a cloudy sky. . . ." But this description is surely applicable to any time of drought and dust storms. Moreover, a few pages later Velikovsky himself cites a quotation to the effect that "references to foreign invaders, to the scanty Nile and to a veiled or eclipsed sun" are "much of the characteristic stock-in-trade of the Egyptian prophet." In other words, such events were frequently foretold, and there is no reason in-

1. Forrest [109] has recently compared, in much detail, Velikovsky's interpretations with the original sources, seeking to demonstrate that these interpretations are far-fetched or unwarranted in many instances.

herent in the description to interpret any such prophecy as describing the particular events with which Velikovsky is concerned.

On pp. 49 and 50 Velikovsky recounts discussion of a question of which the papyrologists said, "It is doubtless wisest to leave the question open for the present." But, undaunted and without adducing any new evidence, Velikovsky proceeds dogmatically to answer the question.

Concerning Arabian traditions (p. 63), Velikovsky refers to authors of the ninth to fourteenth centuries for "narratives which did not have their source in the Bible or the Haggada. They must have been autochthonous and transmitted from generation to generation on the Arabian peninsula." How can one be so sure that over a period of between 24 and 29 centuries (since the catastrophes) these stories were transmitted faithfully and were not influenced by contacts with other cultures?

On p. 69 Velikovsky guesses at what the Hebrew rendering of "sending of evil angels" might have been; finds that it differs by only "one silent letter" from the Hebrew for "invasion of king-shepherds"; and concludes that "the second reading is the original." To most of us, this would be a risky speculation at best; to Velikovsky, it is a compelling interpretation. On p. 132 he gives another lesson in etymology. The usual attempt to relate the word "pontifex" to "bridge" (Latin *pons, pontis*) Velikovsky finds strained: "The word . . . is not of Latin origin. . . . [It is] derived . . . probably from Punt." As support for that, we have only Velikovsky's idiosyncratic interpretation of Queen Hatshepsut's presumed voyage to Jerusalem.

According to Velikovsky's chronology, Thutmose III was contemporaneous with the kingom of Judah. Therefore, he proposes (p. 153) "a new field for scholarly inquiry: the examination of the list of the Palestinian cities of Thutmose III, comparing their names with the names of the cities in the kingdom of Judah. The work will be fruitful." That is guaranteed, because Velikovsky is convinced that he is right, before the detailed evidence is in. Similarly, on p. 156, "exhaustive identification of objects pictured in the Karnak temple and of those described in the Books of Kings and Chronicles is a matter for prolonged study and should preferably be done with the help of molds from the bas-reliefs at Karnak. . . ." Even pending that, Velikovsky's ensuing comparison "will demonstrate the identity" of those objects; Velikovsky has the answer before the evidence has been fully marshaled even by himself.

I found amusing the following illustration of an assertion that something is meaningful, when in fact it proves absolutely nothing. Velikovsky attempts to show that at one time the calendar was divided into 12 months of 30 days each, the whole year being 360 days in duration: "The story of the Flood, as given in Genesis, reckons in months of thirty days; it says that one hundred and fifty days passed between the seventeenth day of the second month and the seventeenth day of the seventh month . . ." [408:335]. Very impressive—unless one does a quick calculation and finds that 150 days also pass between 17 February and 17 July according to our present calendar (except in leap years). More seriously: in the same discussion Velikovsky mentions in a footnote that "the other variant of the story of the Flood . . . has the Deluge lasting 40 days instead of 150." Typically, Velikovsky does not comment on this discrepancy, or produce evidence that one variant is more authoritative than the other or that either of them is reliable. As so often elsewhere, he merely uses whatever fits his own scheme of things and ignores the rest, though guarding himself (if one can call it that) against a possible charge that he does not know the literature. If anyone asks, why not the other variant? Velikovsky can always say that he mentioned it in a footnote.

"In the *Vedas* the planet Venus is compared to a bull . . ." [408:179]. But Indologists are not sure that any planets are even mentioned in the *Vedas*. In the passage quoted by Velikovsky (with an inexact reference) it is Prajapati who is compared to a bull, Prajapati being a well-known god "whom no one before Dr. Velikovsky has ever thought of identifying with the planet Venus . . ." [95].

Velikovsky speaks with the voice of authority even when the subject is not his own work. Thus the Dark Ages of Greece are a figment of the imagination of conventional historians, because "a literate people cannot forfeit completely a well-developed literacy . . ." [431]—it is as simple as that. Geologists are wrong to have accepted the theory of continental drift, which is merely "an intricately contrived *ad hoc* explanation in support of Lyell's uniformity . . ." [166]. Paleontologists are wrong, too: "the theory that fossils are formed when animals die in shallow water is 'impossible.' (Who ever saw a cat wading in water?)" [398]. Those who use the technique of radiocarbon dating also need to be instructed by Velikovsky: "I could not and should not satisfy myself with this support [of his chronology by radiocarbon data] without repaying by demonstrating where the difficulties and pitfalls of the method are hidden" [438]. And Velikovsky has to instruct in thermodynamics the astronomers who contemplate "the completely unsupportable hypothesis of a greenhouse effect as the cause of Venus' heat, even in violation of the Second Law of Thermodynamics" [445]. Only Velikovsky's own work, in his view, is really convincing: "whoever reads it [*Earth in Upheaval*] cannot remain a believer in slow evolution." In 1972 he said, "I would like to be confronted with a scientist who has read my books (not just glanced at them or read the reviews) and then offered me a straight refusal of belief. I have not seen him" [398].

Velikovsky habitually referred to books that he had ready—or almost ready—for publication, which he cited as evidence, and yet which never seemed to see the light of the actual day of publication (see Chapter 5, *Velikovsky's Unpublished Works*). This caused concern even to some of his strongest supporters: "The continuing non-publication of major portions of Velikovsky's research . . . has become, after two decades, a serious damper to all discussion. A panelist at the Duquesne symposium was forced to ask his listeners 'to accept as basic premises . . . five hypotheses put forth by Velikovsky.' . . . These five hypotheses were drawn from the unpublished work. How many scholars will enter such discussions while the detailed evidences remain hidden in galley proofs and dust-gathering manuscripts?" [312].

The answer, of course, is that no scholar—indeed, no person with any sense—is going to engage in serious discussion with one who gives as evidence something that has not been published and is not generally available. It tells something about Velikovsky: that he expected others to take his word that he had the facts, and that those facts supported him; that in good time he would release the data, but until then they should be taken on faith. How could Velikovsky have expected reasoned discussion of his cosmic scenario when he presented only the last part of it? Apparently the original manuscript of *Worlds in Collision* contained "Velikovsky's treatise on earlier catastrophes in the solar system. . . . The entire series of catastrophic encounters among the planets, in his view, constitutes a chain of events cascading down from one initial accident. But . . . scientists and scholars have been asked to accept the history of several terminal events in this series without having a look at what supposedly started it . . ." [306].

Velikovsky's reconstruction of history was long delayed in publication (see Chapter 3, *Ages in Chaos,* and Chapter 5, *Velikovsky's Unpublished Works*). In considering these delays stretching over more than 30 years, it is worth noting that the outline of this reconstruction had already been privately published, in 1945, as *Theses for the Reconstruction of Ancient History from the End of the Middle Kingdom to the Advent of Alexander the Great* [406]. Velikovsky had not changed his ideas in the meantime—he had simply been collating material to support them. One wonders how he could have been so sure in 1945 that his reconstruction was correct, when it took him decades more to assemble the evidence.

Apart from the two monographs or synopses in *Scripta Academica Hierosolymitana,* which are not readily available, Velikovsky published six books before his death but mentioned more than twice that number by title or content (see Chapter 5, *Velikovsky's Unpublished Works*). Many people think of writing books, but not many people think of an unpublished book as something already accomplished, to the extent of feeling free to refer to its contents and conclusions as proving, demonstrating, even illustrating something. Velikovsky had no such inhibitions. He knew that he was right, and it was only a matter of mechanics to get it all on paper and into print.

Never Wrong

In his view Velikovsky had withstood all criticism: "no argument [raised against *Worlds in Collision*] was left unanswered." He was always right: "no chapter of *Worlds in Collision* needs to be rewritten and no thesis revoked" [410: foreword] and "*Worlds in Collision,* as well as *Earth in Upheaval,* do not require any revisions, whereas all books on terrestrial and celestial sciences of 1950 need complete rewriting . . ." [445].

So convinced was Velikovsky of the correctness of his views that he would not admit, or perhaps did not recognize, some of the points raised as criticisms. Thus, replying to the critical review [269] by Neugebauer: "On one point only is he right: I should not have quoted from Kugler's Babylonian Moon table without questioning the age of the tablet . . ." [430]. But Neugebauer had raised a much more serious issue, that Velikovsky mistranslated from German to support his idea that a change in celestial motions had occurred, whereas the original—quoted in German and also correctly translated by Neugebauer—refers not to changed motions at all but merely to the transferring of observed declinations from one frame of reference to another.

One man who felt that some coal deposits might have been formed, as Velikovsky had suggested, in catastrophic circumstances gave credit to Velikovsky for that suggestion, but ventured a minor criticism: "I cannot accept his fourth premise . . . 'If the debris becomes impregnated with bitumen (i.e., from an external source) the coal formed is bituminous coal.'" To which Velikovsky rejoined, "The term 'bituminous coal' is not of my making—it is well established. . . . Bituminous coal has often more than one percent of bitumen, but even one percent of it needs explanation as to its origin . . ." [294]. Velikovsky's affirmation that the term "bituminous" is not of his making is beside the point. He apparently believed that use of the term implies an external origin of the bitumen, which is not the case. The term is used to describe coals having particular properties, among them that one can, under suitable conditions, extract some bitumen from them—which is not at all the same as saying that the bitumen as such is contained in the coal; it may be formed from precursor materials in the "extraction" process.

It does seem, then, that Velikovsky was utterly convinced that he was right, and even minor criticisms seemed to strike him as an affront. The manner by which he came to such certain knowledge on so many diverse matters is clearly of interest, and there are two different versions of that. In 1973 Velikovsky said that he "spent ten years on this work" [431]; in 1972 he referred to "views carefully arrived at in more than a decade of work . . ." [445]. The implication seems clear that Velikovsky spent ten years collecting, sifting, weighing evidence before finally reaching certainty about his reconstruction of history, the corollary scenario of cosmic events, and the law of gravitation.

On the other hand, Velikovsky has himself revealed that he understood all these things within a few months, at most a few years, of his reinterpretation of the date and significance of the Papyrus Ipuwer, from which all else followed: "It was in the spring of 1940 that I came upon the idea that in the days of the Exodus . . . there occurred a great physical catastrophe. . . . Already in the fall of that same year . . . I felt that I had acquired an understanding of the real nature and extent of that catastrophe . . . " [408:preface]. He had moved to New York "in 1939. Within three years he had begun circulating for comment manuscripts that would eventually blossom . . . as *Worlds in Collision, Ages in Chaos,* and *Earth in Upheaval*" [117]. The first draft of *Ages in Chaos* in fact dates from 1942 [301]. On another occasion Velikovsky claimed an even faster reaching of conclusions: "The idea of a cosmic catastrophe in historical times came to me one evening in October 1940: it was inspired by the chapter in the Book of Joshua where it is told about the stasis of the Sun and Moon, and the stones that fell from the sky. In a few weeks the major part of the theory presented in *Worlds in Collision* was conceived . . ." [451].

Velikovsky mentioned how long ago he had reached his understandings in the context of demonstrating the originality of his ideas and his priority over Schaeffer, for example, who reached somewhat similar conclusions about catastrophes in the Mediterranean [431]. Intense concern that he be given credit for being first was quite characteristic of Velikovsky; he seems to have copyrighted his ideas on a grand scale: "On January 28, 1945, I registered a lecture copyright titled 'Transmutation of Oxygen into Sulfur' . . . before the fission (atom) bomb was dropped" [425]—did Velikovsky not know that the transmutation of elements had been demonstrated decades earlier? "A direct experiment to measure the velocity of light relative to moving bodies was devised by me as long ago as 1944, and in December of 1945 it was submitted to the National Academy of Science . . . (also copyrighted as lecture on February 23, 1945) . . ." [441]. "In 1945 or 1946 I registered a lecture copyright on 'Neon and Argon in Mars' Atmosphere' . . ." [452].

Yet, with all that evident concern to be given credit and priority for his ideas, Velikovsky saw himself as someone above such petty concerns: "After all, it really does not matter so much what Velikovsky's role is in the scientific revolution that goes now across all fields from astronomy, with emphasis on charges, plasmas and fields, to zoology with its study of violence in man . . ." [445].

Self-Image

Velikovsky saw himself as judicious and calm while his opponents were not; others used violent rhetoric, but not he: "Believing that an emotional atmosphere is not well suited to fruitful debate, I have entered only infrequently into the controversy" [431], and used "never . . . a harsh word against those with whom . . . [I] disagreed" [445]. He was above the verbal battle, refraining from polemic and the casting of aspersions at his critics. But my reading of the record says otherwise. As far as rhetoric and polemic are concerned, Velikovsky gave as good as he got. Of his critics, he said the following (see also references 272, 303, 416:183, 416:222, 426, and 429):

> The first rule of the scientific attitude is to study, then to think, and then to express an opinion. A reverse of this is . . . what has been done by a group of scientists who have expressed opinions about my work. [35]

> Scientists can calculate the torsion of a skyscraper at the wing-beat of a bird, or 155 motions of the Moon and 500 smaller ones in addition. They move in academic garb and sing logarithms. They say, "The sky is ours," like priests in charge of heaven. [430]

> Dull ears and dimmed eyes will deny this evidence, and the dimmer the vision, the louder and more insistent will be the voices of protestation. [411:preface]

> . . . those who stopped thinking since graduating. . . . [445]

Through Velikovsky's Eyes

Velikovsky's self-centered view of science, history, and other fields is also revealed in statements that are misleading, if not downright wrong. For example, the extent to which established scientific theories are held too dogmatically and for a longer time than might be warranted has been much discussed by historians and philosophers of science, and even by scientists themselves. The concern has been to explain and analyze the *mixture* of open- and closed-mindedness that is characteristic of scientific activity. Velikovsky, however, saw only one side of that coin: "Science today, as religion in the past, has become dogmatic. . . . A scientist must swear loyalty to the established dogmas . . ." [35]; "we are told that the fundamentals are all known" [431].

Velikovsky asserted that, as a matter of principle and dogma, astronomers exclude electromagnetic phenomena from their considerations (see also Chapter 7 for other instances of this): "The reluctance to recognize the existence of electrical and magnetic forces in the celestial sphere . . . is in danger of becoming a dogma . . . to protect existing teachings in celestial mechanics" [429]. In Newton's time the theory of electromagnetism had not yet been worked out: "Would you listen to anybody discuss the mechanics of the spheres who does not know the elementary physical forces existing in nature? But this is the position adopted by astronomers who acclaim as infallible a celestial mechanics conceived in the 1660s in which electricity and magnetism play not the slightest role" [431]. He also referred to "the concept—basic in science until very recently—that gravitation and inertia are the only forces in action in the celestial sphere" [417]. In fact, to this day astronomers calculate celestial motions to the same accuracy as that with which observations can be made, without having to consider electromagnetic effects. They do so not on dogmatic grounds but because all effects other than gravitational ones are evidently so small as to give rise to no observable phenomena in the movements of the planets. Astronomers neglect electromagnetism for pragmatic and empirical, not dogmatic, reasons. Velikovsky's description of this neglect as "dogmatic" is misleading, and it is a red herring.

Similarly, Velikovsky asserted that scientists hold dogmatically to some uniformitarian doctrine, which he set up as a straw man and then demolished: "the principle of uniformity, or the explanation of all past events . . . in terms of the processes in action in our own age, or the denial of catastrophic crises in the past" [431]; "[the] principle of uniformity, according to which only those processes that are observable in our time could have taken place in the past. . . . This principle, however, is made to a veritable bed of Procrustes if 'in our time' is reduced to what an observer can witness within the confines of his lifetime . . ." [435]. That is at best a caricature of contemporary "uniformitarian" views. Explanations are sought in terms of natural *forces* known to be active now and therefore assumed to have been in action earlier, a very different thing from invoking only *processes* now going on. The world of science uses Occam's razor: explanations use the simplest possible hypothesis. If one can explain present and past events in terms of the same forces, one prefers to do that than to invoke different types of events and actions in the past. So Velikovsky's definition of uniformitarianism as the doctrine of contemporary scientists is a straw man. But then, when evidently scientists do *not* cling to that "dogma," it is not because they have more sense than to do so but because of Velikovsky's influence on them: "Today many of the prominent evolutionists of the Fifties have been forced to become catastrophists, but call themselves 'neo-catastrophists' to separate themselves from me . . ." [398]—again that self-importance.

We are told that "the electromagnetic nature of the universe . . . [was] deduced in *Worlds in Collision* from a series of historical phenomena . . ." [431]. It becomes a discovery made by Velikovsky—ignoring all the subspecialties in science that deal with electromagnetic phenomena. Velikovsky did not "deduce" "the electromagnetic nature of the universe"—he postulated that electrical and magnetic effects of an unspecified nature occurred in postulated planetary encounters; that is a very different sort of thing.

In the preface to *Earth in Upheaval* Velikovsky referred to the evidence of his referenced "geological and paleontological material . . . as with the pages of the Old Testament or of the *Iliad*, nothing can be changed in it." The implication is that his critics would wish to make such changes. What the critics quarrel with are Velikovsky's interpretations; he implied, misleadingly, that they do not accept facts.

In referring to his successful advance claims, Velikovsky said that he had included in them that Venus has a "massive atmosphere" and "abnormal (disturbed) rotation" [424, 445].

Such claims are not to be found in *Worlds in Collision.* He claimed hydrocarbons in the atmosphere but said nothing about the density or extent of the whole atmosphere itself. Although Velikovsky could rightly have said that retrograde or anomalous rotation of Venus is consistent with his hypothesis, that is not the same as claiming that he had specifically predicted it. Moreover, it is in itself misleading to refer to the rotation as "disturbed"—no evidence has been presented that Venus's motion ever was other than it now is; the observed retrograde motion simply means that Venus rotates in the opposite direction from most of the other planets and their satellites.

Again, Velikovsky invited misconception when he said, about *Worlds in Collision,* "The reader . . . is invited to consider for himself whether he is reading a book of fiction or nonfiction . . . invention or historical fact . . ." [408:preface]. There are, after all, many other possibilities: an attempt at nonfiction that so deviates from reality that a harsh critic might call it fiction, or nonfiction that contains many errors of fact or interpretation. Even if all the "historical fact" is indeed accurate, the correlations could be completely fallacious. Velikovsky attempted to limit the available choices in the reader's mind, to steer the reader's thought in a direction more likely to produce a result that favors Velikovsky.

"In a few cases it is impossible to say with certainty whether a record or a tradition refers to one or another catastrophe. . . . In the final analysis, however, it is not so essential to segregate definitively the records of single world catastrophes. More important . . . is to establish (1) that there were physical upheavals of a global character in historical times; (2) that these . . . were caused by extraterrestrial agents . . ." [408:preface]. Not so: Velikovsky's evidence supports his scenario only if the records do in fact refer to the same times. If the long night of the Mesoamerican tradition dates from some quite other time than that of Joshua's miracle, then the tradition is no evidence for him, perhaps even negative evidence, and it should have been omitted or, preferably, argued against. The whole thrust of Velikovsky's reasoning is that the same events are referred to in records and traditions around the world, and if these events were not synchronous, then the whole scheme collapses. One might then explain the catastrophes as local ones—earthquakes, volcanos with concomitant darkness as the erupted ash and dust hang in the air, and so on. Since the majority of tales reported by Velikovsky cannot be ascribed in origin to definite dates, he had good reason to attempt to argue that those dates are not decisive. They are, however, quite decisive, and his attempt to portray them as otherwise is an attempt to mislead.

A similar ploy appears in *Ages in Chaos:* ". . . archaeological, chronological, paleographic . . . details are . . . necessary to establish the main thesis of this work. Therefore, any attempt to read this book cursorily will prove to be a fruitless undertaking"; on the other hand, "I claim the right to fallibility in details. . . . However . . . the critic should carefully weigh his argument against the whole scheme . . ." [410:foreword]. Thus the case is built on, arises out of, the details, but it should not be thought to collapse if the details are found to be fallacious. Velikovsky would have it both ways if he could, and argues his correctness whether or not the details bear him out. He will accuse his critics of nitpicking if they point out a few errors and thus demands the impossible—that he be proved wrong only by showing that each and every one of his interpretations and other details are fallacious or spurious. But the onus of proof is on him to show that the case as a whole is convincing, and if the details on which he builds it are not in themselves convincing, then the case as a whole cannot be.

Even the very fact that his ideas were not accepted became, in Velikovsky's eyes and on his chosen ground, an argument for accepting them: "If there is no truth in the work, why is so much ammunition directed against it?" [429]. Well, perhaps simply because it is wrong but was launched with so much laudatory publicity; that might well serve to infuriate a number of people.

". . . Never in the history of science has a spurious book

aroused a storm of anger among members of scientific bodies . . ." [410:foreword]. Even if that generalization holds true for the past, it does not prove that such a book now or in the future is not spurious or, to use plainer language, wrong, misguided, fallacious. Here Velikovsky is employing the Fulton *non sequitur,* and one can turn against him the argument he used when critics said that comets were much smaller than Venus: the mere fact that this has not yet happened, or been known to happen, does not mean that it is impossible.

A historian who discussed and criticized Velikovsky's chronology [386] was admonished: "The very foundations of Stiebing's arguments are not well thought through, for they are based on familiar text-books, whereas I in my reconstruction disagree with the chronology used by these books" [442]. So, quite explicitly here, Velikovsky states that those who wish to argue details with him must do so on his own ground, accepting such basic premises as the correctness of his revised chronology.

Throughout Velikovsky's writing runs the subtly misleading attitude that the onus is on his critics to prove him wrong. Whenever he states—as he often does—that his case is unshaken and has not been disproved, the unwary listener or reader is led to expect that some clear disproof is called for, and that in its absence Velikovsky's reconstruction stands as plausible or even valid. But in all fields of knowledge the onus of proof rests on the new proposition. Progress has come through building on what is known, modifying theories when the established ones have become clearly inadequate, recognizing that anomalies and unexplained phenomena will always exist because no certain and final knowledge is achievable. Velikovsky did not accept that mode of procedure. Rather than remaining open-minded about the significance of parallel descriptions in the Papyrus Ipuwer and in the Bible, he discarded what generations of scholars have built—detail upon detail—into a chronology of the Middle East. Rather than remaining open-minded about a literal interpretation of legends concerning planetary gods, Velikovsky discarded the whole successful framework of astronomical calculations. Throughout, Velikovsky did not accept that the burden was on him to show and to prove, in detail, the inadequacy of the conventional views and theories and, further, that he had a better scheme to put in its place. Taking only what favored his ideas, he constructed a new chronology and a "hand-waving" new physics and astronomy, in which qualitative and poetic descriptions replace quantitative calculations. Then he said, in effect, "Prove me wrong; and do it on ground of my choosing, by criteria I select."

Signs of the Crank

The characteristics illustrated above are consistent with the notion of "crank." He does not accept the methodologies of the fields about which he writes; in at least some, he displays a considerable ignorance, of which he remains unaware. He does not accept the onus of proof, but makes his assertions and insists that they be accepted unless they can be proved wrong to his satisfaction. He is convinced that his work is of signal importance, with ramifications in all areas of thought, and in those areas he rejects accepted views and proclaims his own. He is a universal scholar, a polymath, surrounded by misguided specialists; he corrects them on technical details even in their own specialties. So very important is all this that any humor, any lighter touch, is quite out of place; after all, he walks in the company of Maxwell, Roentgen, Bruno, Einstein, and the rest. Even criticism of quite minor points in his scheme is unacceptable; his view must always prevail. He sees himself as calmly objective when in fact he indulges in polemic and counterpolemic, is quite subjective in his judgments, and from the beginning presents himself aggressively (as well as grandiloquently) as a heretic. He demands credence for work that he has not yet published. Opposition is a mark of conspiracy against him. Every discovery and every public controversy that bear at all on any of his views are seen only in that light: intellectual activity consists of a struggle between Velikovskian and anti-Velikovskian ideas, even when no one but himself has brought his name into the matter. The views of the conventional scholars are based on dogma, whereas he has deduced the truth empirically, *ab initio,* from facts and phenomena.

That, I believe, is an accurate summary of the traits shown by Velikovsky and illustrated with a few examples in this chapter. Yet, because of his unquestioned erudition and mastery of argument and implication, and because he fought the battle so determinedly on ground of his own choosing, he succeeded to the extent that even many who disagreed with him totally nevertheless hesitated to label him "crank" [117, 127, 260, 370]. And, of course, not a few accepted him at his own valuation. My purpose is not to quarrel with these others. My aim has been to document my own attempt, begun initially without conscious prejudice, to reach a conclusion on the basis of the facts, a conclusion with which I could be comfortable, since I had arrived at it myself. To my satisfaction, Velikovsky was a crank.

I am aware that I may be wrong (though, obviously, I regard it as unlikely). Beyond that, I am very much aware that it is possible for a crank to be right and the others wrong. The characteristics illustrated in this chapter describe someone whom I would always be reluctant to believe, one whose logic is foreign to me and whose mode of argument I dislike. But I realize that one can find these crankish characteristics in many others also. Some great scientists have been without humor, full of self-importance, quite contemptuous of opposing views, right by intuition rather than as a result of long inductive wading in facts. And we are all a little paranoid, certainly sure that we are right, unaware of the true extent of our ignorance in some areas. Few of us readily concede victory to the other side in an argument, and in those arguments we also use whatever talents we have for sarcasm, implication, innuendo, the scoring of debating points.

We all partake of those human characteristics. But when we find that someone possesses—in our view—many of these in a marked degree, we single him out as different and label him "crank." It would behoove us in these cases to insert "probably" and leave the question just a little bit open. So, in the present case, I find that, having rejected Velikovsky as a crank as far as the natural sciences are concerned, and having found in him personal characteristics that seem consonant with that conclusion, I am nevertheless left with a nagging curiosity whether some of his many suggestions about the history of the Middle East might not be fundamentally correct. I remind myself that there is some truth in much of what he says, about the dogmatism of scientists, for example. My real objection is to the extremeness, the one-sidedness, of Velikovsky's assertions, and the judgment of what is *too* extreme, *too* one-sided, is purely a subjective one.

10

Intermezzo

In the last four chapters I have outlined the train of thought that led me at last to feel satisfied that I had reached a reasonably objective view of Velikovsky and his work—"objective" in the sense that I began with no preconceptions (other than those inherent in being a scientist) and can point to facts to support my opinions. I summarize those opinions below. Having reached that stage, however, inevitably I was intrigued by questions more far-reaching than "Is Velikovsky right?" I had answered that question to my own satisfaction in about a year of (admittedly intensive) reading and thinking, yet after some three decades the public debate was still very much alive. Why had it persisted?

In the last part of this book I suggest answers to that question:

1. There are many possible reasons for wanting to believe Velikovsky (Chapter 11).

2. Many factors make it relatively easy to sustain such a belief—for example, uncritical commentaries on Velikovsky's books and on the controversy; latent hostility toward established science; confusion about what interdisciplinary work involves (Chapter 12); gross ineptness on the part of Velikovsky's critics (Chapter 13); not least, the tendentious manner in which the debate has been carried on by all the participants (Chapter 14).

3. Gross misconceptions about what science is, how it works, how valid its truths are. I illustrate that in Chapter 15

in the framework of a discussion of some realities about science.

In the final chapter I try to show that the Velikovsky affair can usefully be seen as a type-specimen of public controversies about technical issues, and I suggest that an understanding of the Velikovsky affair can stand one in good stead when considering many other publicly debated, unresolved questions.

Velikovsky's Work

Insofar as Velikovsky sought to contribute to the natural sciences, he failed. Even though his predictions or claims of specific facts and phenomena may be correct, the framework of his concepts is without use to practicing scientists. There is no set of theories that can challenge the existing ones; there is but an inferred set of events, which may or may not be explicable under existing theories. If those events actually occurred, their explication still awaits work within the disciplines. Velikovsky's attempt to discard the theory of gravity is simply not on.

Velikovsky pointed to a great many things that are not fully understood, and he showed conventional views on a number of matters to be on shaky ground—for instance, Egyptian and Mesoamerican chronology. But his contribution stops at that point. He did not propose acceptable ways of modifying the presently held ideas so as to make them sounder, more reliable, and better able to handle all the facts. His method was subjective: selecting data from among a much larger mass of available information and making interpretations that are singularly his own in many instances. The method is not one that could be commended to others, no matter what successes it might have led to for Velikovsky himself. No community of workers using that type of method could reach a single set of conclusions; no science could be built in this fashion.

The greatest claims made for Velikovsky's work—apart

from the possibility that it is in the main correct—are that it is strikingly original and that it is a model of interdisciplinary effort. Neither of these contentions seems to me supportable (see Chapter 13, *Velikovsky's Originality;* Chapter 12, *Interdisciplinary;* and Chapter 15, *Interdisciplinary Science*). What remains, then, is the palpable fact that Velikovsky's work has been successful in stimulating thought along many lines among a number of people. That in itself is no mean achievement. Velikovsky may have been quite wrong—I certainly believe that he was wrong-headed—but his work has had a wider impact, some of it doubtless beneficial, than does the work that most of us do.

Velikovsky the Man

During my inquiry there occurred to me three questions about the relation between the man and his work. Was it by chance that Velikovsky set out in an entirely new and very ambitious direction when he was in his forties? How much does his method owe to his years of psychoanalytic work?[1] Were not some of his conclusions particularly appealing to him because of his strong identification with Jewry?[2]

The phenomenon of the man in early middle age who seeks to start afresh is a well-recognized one. Most of us have, by then, gone about as far in our professions as we are ever likely to; our children have been born and nurtured to the point that they are reasonably able to be independent—and make little secret of their desire to be so; there is, in the ordinary way of things, not much to anticipate in the form of new challenges and new excitements. Some of us attempt to change careers; some take mistresses; some go the whole hog and divorce, remarry, and start a new family. Perhaps Velikovsky, like so many of us, was ready to take up something

1. Discussed also by Fair [106:159–60, 183].

2. This question has also been raised by Gardner [112, 113:32–35] and by Asimov [17].

quite new, to throw himself with full energy and conviction into a venture that offered new insights and the possibility of new achievements.

I have made plain that I see Velikovsky's method as a purely subjective one; he judged the validity of his interpretations by how satisfying they seemed to him, not to anybody else. Is that not the test of validity used by psychoanalysts? The structure of psychoanalytic theory is largely fixed and taken on faith from the pioneers, Freud in particular. The analyst has built that body of theory into himself and needs the help of no one else when interpreting specifics: he has the theory, he alone (apart from the client) has the specific facts, and he himself makes the necessary connections and interpretations. The latter are also judged by the client in an entirely subjective way: if catharsis follows upon insight, that insight is thereby regarded as validated, as objectively true.

Further, there is the approach to the use of data. By the very nature of the psychoanalytic process, practitioners have very few data upon which to construct generalizations—the process of analysis is long, and each analyst experiences a relatively small number of clients. Psychoanalytic data are too sparse to be lightly discarded, and it is a temptation to build generalizations upon singular events, particularly since the validity of these events is not questioned. Is there not some similarity between that approach and a willingness to build a whole new cosmology on the basis of a single, lengthy, subjective chain of reasoning?

The commonly quoted biographical snippets about Velikovsky make clear that being a Jew had considerable practical significance for him, and that he felt a strong identification with the Jewish heritage:

> . . . His father . . . was a businessman and a Hebrew scholar. . . . Because he was a Jew, Velikovsky was unable to attend a Russian university; and he went to Montpelier . . . [where] he organized and became the leader of a group of Russian Jewish students, many of whom became Zionists. Seized by an irresistible urge to go to Palestine, he discontinued his studies to go there. . . .

> Velikovsky . . . traveled to Berlin in hopes of helping to establish a Jewish University. Together with Dr. Heinrich Loewe, a noted German Jewish scholar, he published the *Scripta Universitatis,* a series of volumes containing articles by outstanding Jewish scholars throughout the world. . . . Chaim Weizmann . . . President of the World Zionist Organization . . . asked Velikovsky to organize the Hebrew University in Palestine. . . .
>
> . . . Velikovsky . . . journeyed to Palestine in 1924 . . . [later he] planned the establishment of an academy of science in Jerusalem and started a new series, *Scripta Academica Hierosolymitana*. . . . [472]

Few Russian Jews have not experienced active anti-Semitism personally. Few Jews around the world did not concern themselves with what was going on in Hitler's Reich from 1933 to 1945, and the necessity that Hitler should not win World War II was evident to all. In the preface to *Worlds in Collision* Velikovsky recalls that it was "in the spring of 1940 that I came upon the idea that in the days of the Exodus . . . there occurred a great physical catastrophe. . . ." Among the conclusions that he reached within the next few years were these:

> It was the lot of Saul to carry on the war of liberation of Israel [against the Amu-Hyksos-Amalekite oppressors]. . . . That the Israelites were not remembered with praise for what they had done for Egypt and were referred to as "one" and "they" in Egyptian history was the least of the injustices; their real reward at the hands of the Egyptian historians was to be identified with the ravagers [Hyksos] whom the Israelites had driven out. . . .
>
> Manetho, many centuries later, wrote that . . . the Solymites (the people of Jerusalem) . . . conquered Egypt [and] . . . were extremely cruel to the population. . . .
>
> This confused story reflects the Assyrian conquest of Egypt. . . . The people of Jerusalem never conquered Egypt.
>
> In the Greek world no signs of racial antipathy toward the Jews can be traced until the stories of Manetho began to circulate. . . .
>
> The hatred, never extinguished in the memory of poster-

> ity, for the inhuman shepherd-conquerors was revived: the Jews were identified with the descendants of the Hyksos. Inaugurated by Manetho, an extensive Jew-baiting literature followed. . . .
>
> The Israelites endured much suffering from this distortion of history. They bore their pain for being identified with the Hyksos. The persecution started with the misstatements of Manetho, the Egyptian, whose nation was freed from the Hyksos by the Jews. In later years anti-Semitism has been fed from many other sources. [410:95–98]
>
> . . . Manetho was an Egyptian writer, historian, polemicist, and anti-Semite, inventor of a baseless identification of Moses with Typhon, the evil spirit, and the Israelites with the Hyksos; also, contradicting himself, he identified Moses with the rebellious priest Osarsiph . . . who called to his help the lepers of Jerusalem in his war with his own country. [439]

Thus, during the early part of World War II, when Hitler's Axis was more than holding its own against the Allies, Velikovsky "discovered" that the earliest known anti-Semitism had originated in a baseless historical misidentification. This "discovery" was surely very welcome to a Jew in those unhappy days.

Velikovsky's Critics

I resolved to write the foregoing analysis because I could not find, 25 years after the publication of *Worlds in Collision,* a satisfactory discussion of the merits of Velikovsky's work. That constitutes implicitly a resounding condemnation of Velikovsky's critics. In Chapter 13 I detail explicitly why the criticisms of Velikovsky's work were ineffective or counterproductive.

But Velikovsky's critics were not only ineffective; many of them also behaved offensively. It cannot be gainsaid that literally inexcusable steps were taken to prevent the expression of opinion: a respected publisher was boycotted, individuals were caused to lose their livelihoods, advertisements for books and journals were refused, previously available meeting rooms were withdrawn (Chapter 2, *Conspiracy and Skulduggery;* Chapter 4, *Something Amiss in Science;* Chapter 5, *Something New, Something Old*). Many of Velikovsky's critics argued tendentiously and untruthfully (Chapter 14, *Word Games* and *The Numbers Racket*) while accusing Velikovskians of doing just that: acts of intellectual dishonesty on the part of the critics. As in similar controversies (Chapter 8), the bulk of the criticism was dogmatic labeling and name calling, not the reasoned discourse in which scientists claim to partake when ideas are under discussion.

Scientists and other expert professionals are increasingly drawn into public debate on issues related to their expertise: the safety of nuclear systems, the desirability of radical medical procedures, the implications of basic research on such matters as genetic engineering, and so forth. As with all professions, that of science must recognize that it is accountable to the wider society, and scientists who act as spokesmen for the profession must learn to behave responsibly—out of a disinterested sense of duty as well as to self-interestedly gain public respect for their profession. It is my hope that the following chapters will stimulate thought about the challenges facing scientists when they seek to sway public opinion to their side.

PART III

Beyond the Velikovsky Affair

Everything's got a moral, if only you can find it.
—Lewis Carroll

11

Motives for Believing

Man . . . is a being born to believe. —Benjamin Disraeli

Believing where we cannot prove.—Lord Tennyson

The Wish to Believe

Labeling a man or a belief "crank" or "crackpot" implies that the ideas concerned are incredible, obviously misguided. One is then faced with the task of explaining how otherwise intelligent people can be seduced by such fallacious stuff. To attempt an explanation along those lines, however, amounts to tackling the problem in reverse or attempting to answer the wrong question. The more valid and meaningful inquiry must begin by recognizing that it is characteristic of human beings to wish to believe and that, on the whole, we do believe that what we see, hear, read, is true, just so long as there are not definite and severe obstacles to believing.

The child, learning from adults, finds that most of what he is told is in fact true:[1] "don't touch the stove—it is hot and will hurt you," and sooner or later we do touch the stove and do get burned. As life becomes more complicated, we recognize that the truth is sometimes shaded: we detect our parents telling us "white" lies, and we ourselves tell little lies to

1. It has been suggested that such severe emotional disturbance as multiple personality may result when children are raised in environments that lack truthful consistency.

achieve certain ends. But still, on the whole, it is quite safe to believe almost all that we are told. Our teachers and textbooks give us many facts and explanations, most of which turn out to be true when put to the test. Most human beings are trained to grow into good believers.

In adolescence and early adulthood our readiness to believe encounters our desire to comprehend the world, to know what we are and why. We choose among the various philosophies and religions to whose propaganda we are exposed—several major and many minor religions, secular as well as spiritual. Our choice among these is not made with an open mind and an attitude of "Show me; prove it!" Our choice is made because we dearly wish to believe in something, to be sure about what is going on, because we are not prepared to face a lifetime of doubt and uncertainty. So most of us adopt one or more of the available beliefs, depending on what we are exposed to, and just so long as the belief does not strike us as totally implausible, as quite contrary to our past experiences. And, having once embraced a belief, we are prepared to ignore all sorts of evidence that points to the inadequacy of that belief.

Commonly, as adults, we continue to be more likely to believe than to disbelieve; gullibility is much more common than skepticism. A particular individual adopts beliefs whose nature depends on many things: the culture in which he was raised; the narrower environment of community and of family; his level and type of education; and such personal qualities as self-image, degree of self-confidence, level of frustration or satisfaction with his way of life. So, inevitably, we do not all adopt the same beliefs. But then we wonder how it is possible for others to believe differently from ourselves. As already said, that is the wrong question. It comes from looking at reality with a fixedly subjective eye: "I am sure that my belief is right; how can my friend not see that?"

We must begin by recognizing that humans are ready to believe almost anything, *unless* there are very high, well-nigh insurmountable obstacles to belief; the human wish to believe places the onus of proof—or rather disproof—on the unbeliever. That is not what we take to be the norm, say in philosophy or in scientific activity, yet it is the natural human norm. In the present context we should not ask how it is possible that people could give credence to Velikovsky, who is so "cranklike." We must begin by recognizing that many people, for various reasons of background and personal inclination, will be ready to believe such ideas as those of Velikovsky just so long as those ideas are not obviously and grossly impossible. Then we need only establish that those ideas are not, in fact, obviously impossible.

That case has already been made: the many attempts to *prove* Velikovsky wrong have been quite inconclusive, as illustrated in earlier chapters. There were no great barriers to believing Velikovsky, except for the specialists in various fields who recognized his wholesale repudiation of satisfactory and well-established concepts in their own disciplines.

Beyond that, however, the fervor shown by many of Velikovsky's supporters demands some attention. Admittedly, there are no particularly good reasons why many people should not have believed Velikovsky initially, but when they chose to believe, why so ardently? Among many of Velikovsky's supporters a fervor is seen that can aptly be described as religious. In some part that could perhaps stem from a perception that Velikovsky's work provides a basis for a reaffirmation of traditional religions [16, 83, 92H, 112], but chiefly, I think, the fervor is that of converts to a new religion.

I take it for granted here that contemporary society offers fertile ground for the appearance of new cults. Others have built that case in some detail, and their conclusions seem to me cogent enough: neither science nor conventional religion offers

> answers to the questions that human beings have always asked . . . about life and death, time and space, creation and destruction. . . .
>
> . . . the cults, while . . . insubstantial and occasionally eccentric to the point of being purely funny, nevertheless do their level best to fill a serious vacuum. [104:10, 246–47]

> Most potently, the unorthodox theory usually promises people something they want, whereas conventional science seems to close doors for the ordinary guy. People wish to believe that there are easy, sure ways of foretelling the future, surviving death, or confirming the literal truth of the Bible. If a half-plausible-sounding method of fulfilling these fervent desires is offered, millions will flock to it. [51:3–4]

In much of the discussion of new semireligious cults there is a flavor of condescension toward those who come to embrace the ready-made doctrines, the "insubstantial and occasionally eccentric" sets of beliefs. Condescension here is surely uncalled for; there *is* a vacuum, and not so many of us are notably successful in dealing with it. No doubt some in the ranks of Velikovsky's admirers accept his views uncritically and venerate him unreservedly, just as one finds uncritical hero-worshipers among supporters of any system, be it a new cult, a traditional religion, or one of the scientific disciplines. But it is evident that Velikovsky also attracted many people who revel in intellectual activity, who enjoy thinking for themselves, and who would not be expected to be content merely to hang on a master's words.

If we exclude all consideration of the possible validity of Velikovsky's ideas, and think only of the subjects he dealt with, it surely becomes clear that he could stimulate thought in innumerable ways simply because of the many fascinating, important, unanswered questions that he raised. It can be fun to think afresh about man's heritage, myths, folklore, superstitions, and reading another's speculations can serve to start us thinking ourselves. A few disengaged observers [244], and even some who disputed with Velikovsky [385], have pointed out that his work might well serve a useful function purely as a stimulant of new thought and discussion. One can hardly read de Grazia without acknowledging that he has been led to read, study, and think about a prodigious variety of subjects through his association with Velikovsky. My own study of the Velikovsky affair has helped me to new intellectual insights. One of Velikovsky's followers has given a candidly revealing account of his own experience:

> I took it [*Worlds in Collision*] home, and started to read it; and the reaction was instantaneously positive: I was tremendously excited. I found that I would come across an idea on a certain page and suddenly all kinds of connections would start coming into my head.
>
> Pretty soon, I found that as I was preparing my lectures (my speciality is Shakespeare) I suddenly began to see things that I thought had some bearing on Velikovsky. . . . Eventually I began to find that reading Velikovsky simply shed an entirely new, additional or deeper light on my study of literature: in other words, I found that there's more to Shakespeare than meets the eye. . . .
>
> . . . I met Velikovsky, Mullen, de Grazia and several others; and that's really where it all started for me: I became an initiate, so to speak. (One is embarrassed to use such terms about oneself—a "believer"; a "disciple"; a "supporter"— . . . when I say that I came across this man's ideas and suddenly whole new worlds opened—it may sound romantic, but for me it's really true.) . . .
>
> I do feel that Velikovsky has opened up a new way of looking at the world for me: I can no longer look at *anything* the way I did before— . . . politics, religion, literature, all the academic disciplines, the institution of marriage, even. . . . [352]

Attractions of Velikovskian Research

Many of Velikovsky's proponents have professional qualifications, and many hold academic or other professional posts. Velikovskian research, however, is outside the normal disciplines and is published in Velikovskian journals rather than in the usual scholarly ones. Velikovskian research differs from normal disciplinary research in several ways.

In Velikovskian research the ultimate conclusions are known beforehand: they must be consistent with global cataclysms caused by extraterrestrial agents. In conventional research such final conclusions are not known beforehand, because ultimately all theories are candidates for disproof; that makes it much more difficult to decide when a particular inquiry has been pursued to the stage where publication is warranted. If results of experiments or accumulated observations fit nicely into accepted theories, that is not regarded as publishable work unless the precision of the data is exceptional, or the values of the parameters obtained could be useful to others, or for similar reasons. On the other hand, if the data conflict with existing ideas, much effort will be expended in looking for possible sources of error, reasons why the systems studied might be atypical, and so on. Moreover, in conventional research one often does not know beforehand whether a particular investigation can even be brought to a definite conclusion. One may find that the methods are not sufficiently accurate for the purpose, or that one cannot sufficiently control all variables to isolate the phenomenon that one wishes to study. All in all, the research effort per publication is likely to be greater in conventional research than in Velikovskian research; the latter can usually be done more quickly.

The validity of Velikovskian research is judged by how well it fits Velikovsky's scenario and ideas. Validity is much more difficult to establish and judge in conventional research. The reliability of the data needs to be laboriously tested by examining the reliability of the methods used and the reproducibility of the results (or the concordance of observations), which can lead to much time and effort spent on minutiae. That the results fit into accepted schemes is not enough to warrant publication—they must have some additional value, and the methods and results must be shown, independently, to be reliable. In Velikovskian research independent evidence is not essential, which is not to say that it is not welcomed when available. In conventional research independent evidence is everything—an idea is acceptable and will be accepted for publication only in the light of evidence that does not itself depend on the idea or interpretation being proposed.

In Velikovskian research no particular expertise in special techniques is needed. On the other hand, in conventional research one needs to apply refined methods acquired in years of study and practice. Velikovskian research can be adequately carried on by reliance on common-sense interpretations, whereas in conventional research great complexities often need to be unraveled and common sense is not necessarily a reliable guide (see Chapter 15, *Science and Common Sense*). Thus conventional research is the province of the specialist; it requires refined techniques, a deep understanding and knowledge of facts and theories accumulated over a long period of time, and interpretations that are not necessarily obvious to the nonspecialist. Velikovskian research, however, is open to any intelligent and enthusiastic layman—no training or professional qualifications are needed. As Brian Moore [253] pointed out, "Velikovskian interpretation of myth is still in its infancy and the field is so vast that, as is the case with astronomy, it is still possible for the amateur to make a substantial contribution." I would qualify that to make clear that only in certain areas of *observational* astronomy (as in the search for new comets) have amateurs recently made significant contributions.

Quite generally, in any new field of study, intelligent observation and a common-sense approach are adequate qualifications. Velikovskian studies are new, and one needs to learn little before participating actively. One needs only to have an idea about how something might fit into Velikovsky's scheme, and a description of that idea is then likely to be accepted for publication [74, 75, 135, 205, 264, 285, 286, 335, 342, 397, 404, 468]. Not that all Velikovskian papers are of that type; much effort has obviously been expended by some of the authors. But in at least some cases the results could not have been published in conventional journals because of demonstrated deficiencies in specialized knowledge [203] or because the conclusions are too greatly at variance with established ideas [334, 343, 345, 368].

So Velikovskian research is comparatively easy: a result is quickly obtainable, and it is now also relatively easy to have it published. Anyone who has ideas or insights that he regards as novel and interesting would like to communicate them to others, and can derive satisfaction from writing about them. I think that this is one of the real attractions of

Velikovskian research for individuals of an intellectual bent. It must be particularly attractive to people who have not been able to make professional contributions in their own fields—Velikovskian research then provides an outlet for the creative impulse. It could also be attractive to those who, though successful, have experienced and perhaps tired of the difficulties of working within one of the specialized disciplines. After laboring within the confines of the judgment of one's specialist peers (who referee one's submitted papers and review one's requests for research funds), where innumerable hours of preliminary work, checking and cross-checking, writing and rewriting, go into a possible publication—how relaxing to be able to write speculatively, with a broad brush, about the most major questions.

One obtains personal satisfaction from a new idea or insight, and that subjective satisfaction is quite independent of any validity that idea might have. The satisfaction is experienced when the idea comes, and before it has been tested. In conventional research one is robbed of that satisfaction all too frequently when the idea is tested against experience and experiment and found wanting. In Velikovskian research, since the general truth of Velikovsky's hypothesis is taken for granted, any idea that comes to one within that framework is by definition valid, and the satisfaction of having made a new connection is unlikely to be washed away by the cold water of established facts and theories. Quite the contrary: the more spectacular the speculation, the more likely one is to find it praised by one's Velikovskian peers—as, for example, de Grazia's praise [74] of Juergens's speculation about nuclear fusion (see Chapter 12, *Wishful Thinking*).

Another satisfaction in Velikovskian research is that which human beings derive from their avocation in contrast to their profession. We are all prone to the desire to excel at other things than our professional specialty. Being by profession a chemist, I tend to discount my contributions to chemistry on the grounds that, after all, they result expectedly from my competence and training: I *ought* to make such contributions as a matter of course. I *ought* to develop new methods, discover new compounds or new reactions, elucidate previously unknown mechanisms, and so on. I expect it, my colleagues expect it, my chairman and dean expect it, and any satisfaction from success is muted by those expectations. So I dream about doing superbly what no one expects me to do—making the perfect cast with my fly-rod, executing a brilliancy at bridge. The late President Eisenhower had substantial achievements to his credit as a military and political leader, yet he longed for high achievement on the golf course. President Nixon is supposed to have regretted that he had not excelled at football. Prime Minister Heath would have been delighted to win the America's Cup for sailing. There is a general human wish to excel at one's hobby, and the opportunity to make lauded contributions without too much difficulty is offered by Velikovskian research to all those who have an interest in virtually any aspect of man's nature, environment, and history.

The existence of a small but very enthusiastic group of Velikovskians also offers the satisfaction of associating with others who share one's own avocation. It can be exhilarating to discover a group of individuals who are passionately of the same mind, and that satisfaction can be cultivated through the Velikovskian journals, the symposia, the Center for Velikovskian Studies.

The attractions of Velikovskian research, then, are real and considerable. One can have one's subjectively satisfying speculations published and praised. One is included in a select group of aficionados. One has a creative pastime. And one belongs to the wave of the future, one is—like Velikovsky, albeit to a lesser degree—a heretic prophet, a man ahead of his time.

12

Accomplices to Belief

Where there's a will, there's a way . . .

In light of the human readiness to believe, then, there is no occasion for surprise at the following Velikovsky gathered: real satisfactions are enjoyed by his followers and supporters. Moreover, anyone with the slightest predisposition to be open to Velikovsky's appeal had much assistance toward giving him credence—the enthusiastic and uncritical acclaim of *Worlds in Collision* in popularized previews and innumerable reviews; the ineptitude of Velikovsky's critics and the extreme and unethical behavior of some of them; and, of course, the infinite human capacity for wishful thinking.

Many are ready to believe someone who, like Velikovsky, projects a charismatic image; he has important truths to reveal, unlimited confidence in the correctness of his ideas, and a professional background that—to many—bespeaks the highest technical qualifications (not just a medical practitioner, which would be impressive enough, but a psychiatrist, one who—we suspect uncomfortably—knows more about us than we ourselves do). Others, concerned with academic freedom and civil rights, were there to support Velikovsky as soon as it appeared that an effort to impede or prevent publication of his book was under way: it could have been any book. And Velikovsky's attack on conventional science will have struck a spark in those who, for various reasons, had their own quarrel with science: those whose own pet ideas have not been accepted; some humanists and social scientists, whose disciplines have been overshadowed by the physical sciences in such matters as research grants, academic salaries, public acclaim and prestige; those who have sought to work unconventionally, between disciplines, and saw in the possible vindication of Velikovsky a demonstration of the value of interdisciplinary activities. But above all, an initial predisposition to believe draws on the tremendous human potential for wishful thinking.

Wishful Thinking

It is, of course, much easier to detect wishful thinking in others than in ourselves. But we all "thob," or "think wishly" as the habit has been poetically described [165:14]. As a liberal, *mea culpa* for thobbing Richard Nixon guilty before the facts were in; for thinking wishly that "integration" of schools by wholesale busing would rush us into the millennium; for assuming that the laws of economics could be suspended because I want to preserve Alaska untouched, to extract coal from the ground without mines, to have my electricity without causing extinction of the snail darter. I am no holier than thou, but it seems a necessary part of the present endeavor to illustrate the manifold devices by which Velikovskians wishly support their beliefs. For instance:

> We are now in the midst of a scientific revolution equal to that of 1895–1920 . . . the three events which led off this new era . . . were . . . 1. Velikovsky's resynthesis of astronomic events. . . . 2. Reines and Cowan's experimental demonstration of . . . the neutrino. . . . 3. Yang and Lee's prediction of the violation of parity. . . . [94]

> Whereas astronomers are perplexed at the implication of the new picture of the universe as derived from the space probes, Velikovsky has been clear from the very beginning. . . . Velikovsky was confident that this evidence would be found, and it has been found. . . . [378]

> Velikovsky's correct diagnosis over a time span of twenty-seven years, so contrary to conventional theories, can only mean that Velikovsky himself is the foremost scientist of the twentieth century. . . . [465]

So Velikovsky's supporters see him as "among the foremost thinkers of all times" [465] (see also [11, 71, 73, 74, 78:1–4, 168, 173, 264, 316, 468]) and believe that mainstream science is incorporating Velikovskian views. There is a notable lack of any wider perspective. This attitude leads to such things as making much of trifles and refusing to admit that Velikovsky could be in error, or could behave other than ideally, at any time or in any way.

Of course, in a general way it is admitted that Velikovsky is not perfect, as no man could be (Chapter 4, *Right or Wrong Is Not the Issue,* and Chapter 5, *Velikovsky*). But it is rare indeed to find a statement from the Velikovskian side that clearly admits Velikovsky to be in error *in a specific instance.* This unwillingness to admit a specific failing leads to attempts at explaining away what might, to others, seem fairly obviously a mistake on Velikovsky's part. One example is the attempt to play down, indeed to disclaim [72, 212], Velikovsky's clearly expressed belief in the validity of an analogy between the atom and the solar system (Chapter 7, Worlds in Collision *and Later Work*). Again, I have pointed (p. 128) to Velikovsky's confusion of contemporary views about the ages of the universe and of the solar system as calculated from the abundances of the elements. Juergens [178] wanted Velikovsky to have said something other than he actually did: "Velikovsky pointedly asks: . . . 'why do we assume that at creation the heavy elements . . . predominated and not the simplest ones? . . .' The cosmologist will . . . reply: 'We do assume that the heavy atoms have been built from the lighter ones. . . .' But Velikovsky's point—and it's a good one—is that no theorist stops to consider the atomic-fusion possibilities of the electric discharge. . . ." To me, it remains clear that Velikovsky asked why we do not assume something that we do, in fact, assume.

Velikovsky himself seems to have said that he was right all along: "First published in 1950, this book was left unchanged in all subsequent printings . . . nor have any textual changes been made in the paperbound edition. . . . I wished to keep the text . . . unaltered . . . [to] face all subsequent discoveries in the fields it covers or touches upon . . ." [409:preface] (see Chapter 9, *Never Wrong*). Isaac Asimov, among others, derided Velikovsky for his profession of unerring omniscience. Commented one of the Velikovskians [168]: "Asimov took Velikovsky to task for saying 'that not one word of his 1950 book has had to be changed as a result of scientific advances since its publication.' This wasn't an accurate interpretation by Asimov. . . ." To me, that seems a very accurate interpretation (very possibly the least inaccurate of Asimov's various criticisms of Velikovsky). But the explaining away continues: "Actually, Velikovsky doesn't consider himself infallible. . . . Certainly there are passages . . . which he would like to rewrite or reconsider, but there is little to be gained in rewriting an awkward phrase or sentence. Moreover, he has in fact, reconsidered and expanded on several of his thoughts. . . . In the main, he is confident that the seventy-odd advance claims made in *Worlds in Collision* are couched in sufficiently lucid language to eschew any modification. . . ." Does this explanation not say clearly that Velikovsky has in fact been right on all points of substance, which was what Asimov alluded to? Asimov was not criticizing Velikovsky's style of writing, yet that is what Velikovsky is here defended against.

De Grazia [73] seemed to support the claim that Velikovsky had been always right: "no single case of misstated fact was proven in any of the four books of Velikovsky. . . ." The accuracy of this statement could be argued, but resolution of the argument would require agreement—unlikely to be achieved—as to what constitutes a "fact" and what constitutes "proven." I would say that the following is a misstated fact: "a theory that envisages not dissimilar events in the macrocosm—the solar system—[as in the atom] brings the modern concepts of physics to the celestial sphere" [408:preface]. I say this is a misstated fact because the modern concepts of physics view quantum changes of orbits as inherently applying only at the level of atoms and molecules, owing to the very small numerical magnitude of Planck's constant. But de Grazia would not agree, having already given his opinion that Velikovsky was merely drawing here an "incidental analogy" [72]. I would also say that when Velikovsky gave an incorrect translation from a German source, thereby actually and drastically changing the meaning [269], he gave a misstated fact. De Grazia is certainly not ignorant of the details of the Velikovsky affair; I would confidently venture the guess that he is as conversant with the literature as anyone. But he sees the matter through the lenses of his conviction that Velikovsky is basically right, so that a demonstrated mistranslation can be easily put aside as of no consequence.

In the eyes of his supporters Velikovsky is not a man to indulge in immodest comparisons of himself with great people. Thus Jueneman [168]: "Although Velikovsky's supporters have compared him to Galileo, Velikovsky himself has discouraged this. For one thing, Galileo was intimidated by the opposition, and he fumbled the ball four out of five times before making a score. [Is Jueneman implying that this is the reason Velikovsky gives for discouraging the comparison?] . . . At any rate, Velikovsky has never compared himself with anyone but himself, contrary to Asimov's assertions. . . ." Quite clearly, however, Velikovsky did compare himself to (or his situation with that of) Maxwell, Marconi, Roentgen [445]; Einstein, Aristarchus, and the Wright brothers [431]; Faraday [432]; Bruno and Campanella [299]; and Sigmund Freud and Victor Hugo [449]. (In his memoirs, Velikovsky adds Darwin to this list [416:61, 63]).

Stecchini [378] records that Velikovsky obtained the satisfaction of a retraction from a critic; I venture that most readers of that "retraction" [181] would judge it, as I do, a reassertion of the original criticism.

Velikovsky's editorships have been made more of than the traffic can bear: "He founded and edited the *Scripta Universitatis,* a joint work of Jewish scholars out of which grew the University of Jerusalem . . ." [209]. I have pondered more than once how the writer envisaged this remarkable growth. Slightly more prosaically, "he and . . . Loewe founded and published *Scripta Universitatis* . . . conceived as a cornerstone for what would become the University of Jerusalem. . . . Einstein edited the mathematical-physical volume of the *Scripta*" [172].

Velikovskians lavish disproportionate praise not only on the work of Velikovsky but on their colleagues' extensions of it as well. Juergens [178] speculated, "I . . . would predict with some confidence that, once the curtains of *thermo*-nuclear theory [i.e. current theory of nuclear structure and reactions] are drawn aside, electrical engineers will quickly discover that the controlled-fusion reactions they have been seeking in vain for a quarter of a century have actually been within their grasp for at least twice that long. . . ." Kruskal, an applied mathematician at Princeton University, devoted two and a half pages in *Pensée* [203] to discussion of Juergens's work on electrical effects. Apart from the technical details, Kruskal gave a sort of summation: "Mr. Juergens' reply . . . betrays once more (I am sorry I cannot put this more politely) his complete ignorance of even the elementary theory of electrostatics. . . ." But the evaluation by de Grazia, political scientist and supporter of Velikovsky, differs: ". . . Mr. Juergens . . . has written articles that I am convinced will be numbered among the most important of our age. . . . In a modest and incidental remark, Juergens has also suggested that the key to the solution of the urgent problem of nuclear fusion . . . may be in the study and understanding of the interplanetary electrical discharges . . ." [74]. That evaluation by de Grazia, like Juergens's suggestion itself, strikes me as a rather extreme example of wishful thinking.

And again: "'Worlds in Confusion'—an . . . Asimov tirade against Velikovsky, long since refuted in its entirety by progressive scholarship in various disciplines" [200]. That will be news to those who have seen no refutation of some of Asimov's points [16], such as that Velikovsky's description of the perturbation by Jupiter of Lexell's comet is quite wrong, or that, contrary to Velikovsky's statement, there is no reason (in the accepted laws of physics) why the orbital velocity of a satellite must be slower than the velocity of rotation of its parent.

In one germane area Velikovskians characteristically indulge in unbridled wishful thinking: regarding what can be accomplished, and how quickly, by scientific research and development. One example is Juergens's notion that controlled fusion has been within the grasp of electrical engineers for 50 years, implying that it could be achieved in short order if only the right experiments (which, incidentally, Juergens did not specify) were undertaken. Another example, from Stecchini [378]: "There is reasonable ground to hope that the new investigation which takes electric charges and magnetic field into account will . . . succeed in explaining the behavior of comets especially in the proximity of the Sun. . . . Thereafter, the case of planets like Earth and Jupiter, which are surrounded by a magnetosphere and move through the magnetic field permeating the solar system and the plasma winds that sweep through it, will come to quantitative analysis, too." I want to focus on the "reasonable ground" for Stecchini's hope, leaving aside all questions as to whether there are even problems to be solved or discrepancies to be accounted for. Stecchini is talking about nothing less than a complete revolution in the types of laws that are used for explanation and calculation. A whole new system of laws would need to be set up, by examining many specific instances before the laws could be regarded as reliably established. All existing data in astronomy would require reinterpretation and a finding that these data could be accounted for as well by the new laws as by the old ones. And that would be a major task for whole armies of specialists—theoretical astronomers and physicists, experimentalists, mathematicians. In practice, that is a very unreasonable hope. It exemplifies what I think of as a science fiction (SF) approach to science.

I read long ago a lot of SF, and well remember a certain type of story that always seemed to me totally unrealistic. A star-ship is disabled, for example, and apparently doomed.

But in the engineers' quarters there is a young gadgeteer who never bothers much with theory but is a whiz at making things that work. With tin cans and sealing wax and the like he builds a *new type* of gadget, working on the basis of a *new principle* of nature, which makes possible a new type of faster-than-light motion: more economical and simpler than the earlier types of engines. Not only that: he constructs a prototype with available materials, and the prototype not only works, but is powerful enough to get the star-ship safely home.

That SF concept of science and technology is utterly at variance with reality. New principles are not established in one fell swoop; the great inventors labored long, by trial and error, for each of their inventions; lengthy development and refinement are needed to bring the new gadget to the point where it works efficiently and reliably. Think, for example, of the Wankel rotary internal-combustion engine. Merely new technology, not even a new principle, is involved, yet many years of development of this theoretical improvement have still not resulted in a model that competes successfully with conventional engines. As to new principles, how long were electrical phenomena known before electricity was harnessed? And so on and so forth. The SF concept of science takes no account of practical realities and difficulties, especially those always encountered when one does anything for the first time. SF science is wishful thinking, and Velikovsky's followers indulge in it freely.

Thus Mullen [264] calls for work that "simply assumes him [Velikovsky] correct and proceeds to further research." That would be SF science, not the real thing; the real science builds on what is already established in science. Mullen wants quantitative descriptions of "a magnetized solar system in hypothetical catastrophic conditions," as if physics, mathematics, and astronomy could give answers to what is, on the basis of the laws that these disciplines work with, a totally unrealistic situation. How would one carry out those calculations? Mullen seems to think that the techniques and methods of these disciplines can be divorced from the body of theory and fact and applied to some entirely different and imaginary body of theory—how? This is not science, not even a suggestion about science, but unfettered, undisciplined speculation. Mullen asks for experiments on "the old Lamarckian theory of inheritance of acquired characteristics." But these are not widely carried out nowadays because so many past attempts failed to turn up evidence for the validity of Lamarck's ideas; despite systematic experiments, no results were obtained that needed to be explained in a Lamarckian way, and predicted Lamarckian effects were not observed.

Science is cumulative, and learns from the past. The Velikovskians would start afresh in each field, not recognizing that this is impossible, especially when they want the disciplines themselves somehow to use, test, expand on Velikovsky's concepts, which contradict the content of those disciplines. "Inheritance of behavior patterns laid down in catastrophic circumstances might explain . . . bird migration; swarming; acute sensitivity of many species to the subtlest earth tremors, solar eclipses, etc." [264]. No doubt it could; so what? It is easy to think of possible explanations for anything; the difficult thing is to establish that a particular explanation is *likely,* and that it is *more likely* than other possible explanations. The test of what is likely is whether there is a concordance with reliable experience, which is embodied in the existing disciplines. Mullen talks of education "towards multidisciplinary mastery"—an SF concept if ever there was one. Few physicists of mature years would lay claim to a mastery of physics as a whole [85:56, 96–98], let alone to mastery of other disciplines as well. How easy it is to say, how difficult—impossible, in fact—to do.

Scientists do not spray out ideas in shotgun fashion as though expecting others to pick them up and work them through. Velikovskians delight in it: "the house of geology needs to be rebuilt from the cellar up," says Mullen. But perhaps the ultimate in wishful SF thinking is the Velikovskians' view of *Cosmos without Gravitation.* I believe I have shown that opus to be without scientific virtue, yet Juergens solemnly quotes it as a reference in his technical discussions of electromagnetism [175], and Jueneman [166] describes it as "a persuasive cosmology without gravitation."

At Face Value

Many people were prepared to accept Velikovsky at his own valuation, at face value. Velikovsky presented himself with confidence and self-assurance as a man of great learning in a range of intellectual fields, at home with the technical jargon and the body of facts and theories; as one who had labored for ten years to draw his strikingly original, heretical conclusions from a mass of disparate data; as one who, though a heretic, had spoken on equal terms with experts in many areas (and some of those experts agreed with him in part at least, or did not dismiss him outright). His book was full of discussions of technical minutiae about facts, interpretations, translations, whose merits most individuals could not themselves judge in an informed manner.

Admittedly, this persona and its implications were rejected by the overwhelming majority of specialists, who could judge for themselves the implausibility of Velikovsky's views as they impinged on their own areas of expertise. But most readers and reviewers had no personal basis for assessing Velikovsky's work; for them, I submit, it was easier to accept—or at least to remain open—than to reject out of hand. How many "ordinary" people have the self-confidence to say, of someone who attracts great publicity (much of it favorable) and talks in learned technical language, "He is a fraud"? How many of us have the comfortable familiarity with the Bible, the related histories of the Middle East, the terminology of astronomy, to reject a man who presents a massively detailed discussion of those subjects, even though that man seems a bit queer, a bit egocentric, a little too self-important? We may not like him, but we will hardly dismiss him.

So, in many of the public commentaries, one finds Velikovsky's own valuation of himself uncritically passed on to the public. He was referred to as "a physician-philologist [who] had previously made distinguished scholarly contributions to a number of scientific disciplines" [42]; "widely-travelled scholar and scientific investigator" [92E]; "physician, scientist, historian, philosopher" [92A]. In obviously awed tone, we were told that "the Doctor observes that . . ." [217, 218; see also 209, 271, 278, 284]. The misleading (Chapter 9, *Never Wrong*) description of his conclusions as resulting from ten years of work was often repeated [92C, 217, 218, 284]. The work was said to be of very high significance and great originality: a "startling interpretation" [217, 218]; "will have an explosive effect in the world of science" [18]; "its inferences . . . are so great and far reaching as to make the hydrogen bomb . . . seem like child's play" [92A]; "the central thesis . . . is stupendous" [92D]; "a bold and startling thesis" [92G]; "startling, astounding, amazing and entirely revolutionary" [92I]; "daring and original" [92P]; Velikovsky is an "arch-heretic" [233].

Not many of us could resist such a succession of superlatives, particularly when the author ranges over virtually all the fields of learning, in the company of the greatest men of the past and present: "Dr. Velikovsky's work crosses so many of the jurisdictional boundaries of learning that few experts could check it against their own competence . . ." [209]; "did not convince his fellow-scholars. Neither . . . did Pythagoras, Thales, Galileo or Copernicus . . ." [92K]; "the initial rejection of Velikovsky's ideas . . . reminded some observers of Galileo's persecution at the hands of the Inquisition. But, unlike Galileo, Velikovsky would not recant his view under pressure . . ." [453]; "one of the biggest arguments since Einstein" [80]; "he got encouragement from Albert Einstein" [66].

Antiscience

Many people have pointed to a widespread distrust of, even hostility toward, science and scientists. It does appear that some of the support for Velikovsky came from individuals who were ready and eager to attack the scientific community (or, as they put it, "establishment").

Quite evidently, the manner in which many scientists attacked Velikovsky is not to be excused; I concur with many of the criticisms made of such behavior, when expressed as criticism of specific actions of specific individuals: for example, Shapley and the boycott of Macmillan. However, many of the criticisms went beyond the specific, and showed the authors' hostility to science as a whole and to scientists in general [92B, F, J, K, M; 378; 403]: "the scientific establishment . . . commits injustices as a matter of course" [71]. A general hostility was certainly shown by the individual [392] who was "deeply shocked, though not surprised" at the scientists' treatment of Velikovsky.

The existence of this latent hostility provides one source of the Velikovsky phenomenon, one reason why Velikovsky's writings make converts irrespective of the possible validity of the ideas and contentions. Much of the abuse heaped on scientists and on science during the Velikovsky affair is less revealing of dominant themes and nuances of contemporary scientific activity than it is of widespread feelings that there is something seriously wrong with "science": "As a symbol of revolt against scientific and academic orthodoxy, he [Velikovsky] became a popular campus lecturer . . ." [396:33]; "we have always distrusted . . . scientists . . . [as] difficult to understand, dogmatic, faintly unsavory, cold. . . . The holder of unorthodox ideas, on the other hand, presents himself as the champion of the common man and the victim of persecution by the scientific 'establishment.' Immediately, this pose endears him to the millions of men-in-the-street who also feel that they are being persecuted by some sort of establishment . . ." [51:3–4].[1]

Interdisciplinary

Characterization of Velikovsky's work as "interdisciplinary" has been common [18, 73, 78:1–4, 212, 379, 389, 466]:

1. Thus Velikovsky in the preface to *Worlds in Collision:* "This book is written for the instructed and uninstructed alike. No formula and no hieroglyphic will stand in the way of those who set out to read it."

"Velikovsky's early bridge-building across both the 'hard' and 'soft' sciences . . . *interdisciplinary synthesis,* digging just deep enough to get a firm footing" [166]; "Velikovsky is an interdisciplinary scholar of the first magnitude" [167]; "Velikovsky has been scorned for blending the study of astronomy with that of geology, ancient traditions, ancient chronology, and ancient science . . ." [378].

This theme is another red herring (see Chapter 15, *Interdisciplinary Science*). Velikovsky was scorned not for blending but for rejecting wholesale the accepted views in several disciplines; for his ignorance concerning much that he discussed or rejected; for saying that astronomers were wrong about astronomy, geologists about geology, historians about history, physicists about physics. Velikovsky's "challenge to conventional views in science" (Chapter 5, *Velikovsky's Later Work*) was not so much a challenge as a repudiation. He was a "rebel against almost all human learning" [133]: not an interdisciplinarian but an antidisciplinarian.

Those who lauded Velikovsky for his interdisciplinary approach were attracted, I believe, not by any specific demonstrable interdisciplinary validity in his work but because he claims to understand so much as to offer an integrated world-view: a substitute for the religion in which we no longer believe, which claimed the answers to all the really important questions in the confidence that all separate disciplines would arrive eventually at conclusions consonant with that overall world-view. It is that sort of global integration, across *all* disciplines, that is felt by many of us to be sadly lacking nowadays. None of the existing specialties concerns itself with this now that theology is on the defensive; being human, we would like to know—or at least to believe that we know—what the purpose is to it all, why there is life and death, how big and how old the universe is. While Velikovsky did not go so far as to say specifically that he had the answers to those questions, he came close enough for many people: man's aggressiveness, the beginnings of consciousness and of religion, for example, all as an understandable result of major catastrophes. I think that this theme, not necessarily obvious upon cursory reading of his work, has been the attraction for some portion of his supporters. So I see the Velikovsky phenomenon as partly that of a secular religion. Velikovsky's work is not interdisciplinary but antidisciplinary; those who call his work interdisciplinary are attracted by the philosophy, the religion, of something that is *trans*disciplinary: an integrated world-view.

Velikovsky and the Student Movement

I have suggested that part of Velikovsky's appeal is that of a secular religion, offering a perspective on all human activity, involving hostility toward science and its institutions, thriving on uncritical and wishful thinking, accepting at face value what one wants to be so. That combination is very much what John Searle [366:74] discerned in the student movement of the mid-1960s to early 1970s: a seeking of abstract, sacred values; uncompromising hostility toward all traditional institutions; uncritical acceptance of the rhetoric of the student leaders. "The answer to the more general question—why is it easier to believe in mythology rather than facts?—is that where the sacred is concerned, people's perceptions are rigidly shaped by their dramatic categories. In the sacred cause . . . 'oppressed-minority-wins-struggle-for-justice-against-reactionary-authorities' is a standard dramatic category, . . . a device for perceiving events. . . ."

The congruence of Searle's analysis with mine offers at least a speculative reason for the growth of interest in, and support for, Velikovsky on college campuses in the mid-1960s (Chapter 5, *A New Climate of Opinion?*). Searle points out that the student (and nonstudent) radicals were able to enlist even moderate and conservative students and faculty by focusing attention on the repressiveness of administrators, and he describes the determination of the radicals to avoid any defined organizational structure in the movement. It seems to strengthen my thesis, that the first Velikovskian magazine, *Pensée,* was the product of nonleftist students in the early 1970s—here was an anti-establishment cause not tied to the political left. And the founders of *Pensée* exemplified the radicals' extreme distaste for organizational stability [311]: ". . . Founded in 1966 and soon thereafter allowed to lapse for several years, it was revived in 1970 as an unofficial student magazine. . . . Confronting a highly politicized campus press that was, in those burn-the-campus-down days, almost without exception 'far left' or radical, *Pensée* offered a 'rational alternative'—political conservatism. . . ." Now *Pensée* was confronted with the choice of seeking financial soundness or going out of existence:

> . . . In our day the choice between risk and security is rarely a genuine choice; security beckons so seductively that few would think of rejecting her. *Organizational* security . . . should certainly appeal to any magazine editor or publisher. And yet, in the search for the truth, risk may be preferable. . . .
>
> In seeking to avoid such a trap, *Pensée*'s surest course may be to embrace the risk of failure . . . to pursue the hope of madness rather than the safety of mere sanity. . . .
>
> Above all else, *Pensée* is an interdisciplinary journal. A key to understanding what this means may be found in a term which, for the scientist, carries all the negative connotations of religion: *conversion.*
>
> To the unconverted man . . . it is clear . . . that *a set of well known facts rules out X.* . . . [And] for the real plausibility of X to be seen, one must suspend his previous judgment on such a broad range of taken-for-granted and revered propositions that simply to urge the necessary shift of mental vantage point is to earn the appellation "madman."

13

Blundering Critics

> Why beholdest thou the mote that is in thy brother's eye, but considerest not the beam that is in thine own eye?
>
> —St. Matthew

In retrospect, it seems clear that Velikovsky's critics did not grasp the magnitude—or perhaps even the nature—of the task of discrediting Velikovsky in the eyes of the public at large. Academics in general, and perhaps scientists in particular, tend to lead lives that are notably sheltered from contact with other people than fellow academics and the industrial and governmental scientists and administrators who impinge on academia. Scholars rarely have special skill in dealing with situations in which wide publicity is involved, in which controversial issues are publicly debated, and in which they cross swords with individuals who do not play the game by the rules that are accepted in academic circles. An academic myself, I recognize how difficult it is for me to believe, for example, that anyone whom I know personally would be capable of cheating on an examination or forging research data. My knowledge that such things happen is an intellectualized knowledge, and I recognize actual occurrences belatedly and with more incredulity than do, for instance, my friends who are practicing lawyers and spend much of their time in contact with thieves, habitual liars, con men, and so forth.

It may well be that nothing effective could have been

done by scientists, individually or collectively, to discredit Velikovsky, to counter the image of importance, significance, competence, confidence, that he projected and that was parroted in so much of the publicity. Be that as it may, it is not difficult to discern that Velikovsky's critics committed blunder upon blunder, and failed to make a case that would convince disengaged but interested and perceptive observers who were not scientists (or possibly historians). In fact, the critics' efforts might well have caused people to side with Velikovsky: "Chesterton was converted to Christianity by reading the denunciations of the Nineteenth century atheists, reasoning that anything which was attacked for so many, and such diverse and contradictory reasons, must have something good in it. The writings of some of Velikovsky's opponents must have a similar effect on many fair-minded people" [252].

The goal apparently sought by Velikovsky's detractors was an impossible one: to disprove him decisively, absolutely. Within the scientific community something of a consensus generally rules as to what degree of probability or improbability is sufficient to justify taking action—mental or otherwise—on the basis that the probability can be translated into effective certainty. The presence of that consensus dulls the realization, if it exists at all, that absolute certainty is not attainable. So scientists are apt to become helpless when others do not make the same estimates as they do of probability, or are prepared to translate comparatively low probabilities into firmly held beliefs.

Finding that their collective judgment was not being accepted at large, the critics responded not by taking extra pains to clarify their argument but by becoming increasingly impatient and strident. They "argued" dogmatically, *ex cathedra,* by analogy, *ad hominem,* with ridicule, in most every other way than by calm, comprehensive discussion of the relevant points. Thus *ex cathedra:* "Velikovsky's ideas hold about as much water as a well-worn piece of cheesecloth" [12]; "all of its [*Worlds in Collision*] basic contentions are dynamically impossible" [146]. Other, equally dogmatic, statements abound [4, 5, 15, 17, 37, 38, 56, 67, 68, 116, 198, 236, 238, 247, 274, 287, 365, 393, 402]. The tone of *ex cathedra* assurance was laced at times by ludicrous hyperbole: "unquestionably the most outrageous collection of nonsense since the invention of the printing press" [96]; "it is a conservative estimate . . . that the American publishing enterprise has been set back at least twenty-five years with one publication of a senseless piece . . . entitled *Worlds in Collision* . . ." [245].

Argumentation in this fashion was bound to be ineffective, for a number of reasons. In the first place, it could not convince those who prefer to think for themselves, and thereby implicitly insulted the intelligence of the audience at which it was directed. Second, scientists are fond of maintaining that they arrive at their opinions by considering the evidence, not by hewing to dogma; this was said a number of times during the Velikovsky affair. By arguing dogmatically themselves, the scientists were not practicing what they preached—"do as I say, not as I do" came through rather clearly. Third, by arguing from authority, the critics made the matter one wherein the audience was invited to choose sides not on the basis of logical evidence but by deciding which individuals to believe—Velikovsky, Larrabee, Lear, Oursler, Fadiman, etc., on the one hand, or Shapley, Struve, Payne-Gaposchkin, Harrison Brown, etc., on the other.

Not unrelated to argument *ex cathedra* is the device of argument by analogy. Teachers of science often use analogy to express somewhat abstract concepts in a more familiar and concrete way. The interchange between students and teachers in a classroom inevitably has an authoritarian aspect, and here analogies can be useful tools. It is quite a different matter, however, to use analogies in disputes: if the disputing parties cannot come to agreement on the specific controversial issue at hand, they will also disagree on the validity of proposed analogies, and the argument will veer off into total irrelevancy as the question of how apt an analogy is this one or that is bandied about. In the Velikovsky affair a number of inevitably futile attempts were made to show Velikovsky to be wrong on the basis of analogies.

Thus Payne-Gaposchkin [289] pictured Velikovsky's account of cosmic near-collisions as "an extraordinary achievement in a very difficult type of marksmanship—four (or even five) hits in a couple of thousand years, and all (by a lucky chance) at crucial points in the history of Israel. . . ." The term "marksmanship" introduces a totally invalid anthropomorphic analogy. Velikovsky had not claimed that Israel's history was foreordained to become significant and that the catastrophes brought this about; he said that these striking events shaped the later history of Israel [408:379–81], hardly an unlikely thing once the premises are granted.

Stewart [385] illustrated, by an analogy, the difficulty of using technical arguments to disprove Velikovsky: "Suppose a sparrow flutters past a tall building. . . . A person who lacked all experience in numerical reasoning but had intense sentiment for sparrows might argue that air currents from the bird's beating wings had dangerously strained the tower. . . . An engineer could . . . prove . . . the absurdity of the assertion . . . [but] would find it difficult in a book review to . . . explain his proof briefly and in non-technical terms. . . ." Then Stewart outlined briefly some of the technical arguments against Velikovsky. The analogy of the sparrow was intended to lend the weight of authority to these later assertions. Is such a mode of argument likely to convince an intelligent reader or to insult him?

Margolis [236] termed Velikovsky's scenario impossible "for the same kind of reason that you cannot pour two quarts of water into a one-quart jar": well, maybe—but Margolis would need to disprove the scenario itself in precisely that way before he could show the analogy to be valid. Haldane [139] "could write as convincing a book . . . to prove that monkeys had originated from men": he would have to actually write such a book, and then show it to have convinced as many other people as did *Worlds in Collision,* before that analogy would pass muster. Menzel [173] compared Velikovsky with "Dr. Brinkley of the goat-gland era"; Nichols [274] used the analogies of perpetual motion, the sidehill wampus, and so forth. Are these arguments? Certainly not reasoned ones. Perhaps it is even too kind to class them as arguments by analogy; they are arguments *ad hominem,* of which there were many others. Velikovsky was said to belong to the class of "the self-styled scientist, incompetent in his field, but living under a delusion of greatness and driven by unconscious compulsions to create off-trail theories of incredible complexity and ingenuity" [112]. "The seventh footnote on page 280 shows clearly that one of the 'authorities' [Atwater], whose praise of the book is printed on the jacket and whose 'open-minded' (or empty-headed!!!) discussion was given wide circulation in a recent issue of a weekly magazine, is just as big a screwball as Velikovsky" [96].

Arguments *ex cathedra,* by analogy, and *ad hominem* were spiced with ridicule [139, 185, 214, 238, 269, 274, 287, 384, 393, 400–402], as for instance: "times and the treatment of heretics have changed. Bruno was burned to death, and Galileo . . . languished under house arrest. Velikovsky won both publicity and royalties . . ." [127]. Quite naturally, these modes of discussion converted none of Velikovsky's supporters. Nor were they notably effective with the public at large or with such initially disinterested observers as Stove [389] and the author of this book, for example. And the critics made fools of themselves in a number of ways:

By on the one hand saying that *Worlds in Collision* was obviously nonsense and on the other hand expending much time and emotion attacking it, the author, the publisher, and those who did not dismiss it out of hand;

By launching full-scale attacks on the basis of summaries and previews, without having read *Worlds in Collision* itself; even advising others not to read the book while stating that they had not read it themselves [429];

By claiming that our high standard of living proved our scientific knowledge to be sound [274];

By saying that any verification of Velikovsky's predictions meant nothing, "since the idea is wrong" [247];

By asking fellow scientists not to publish work that could be construed to give comfort to Velikovsky [173];

By stating to be impossible changes in planetary orbits that are not impossible [24, 25, 248] according to existing knowledge;

By saying that Velikovsky's "glamorous and sensational theories . . . *can be proved wrong*" [39] and then failing to furnish the proof;

By calling Velikovsky's work so vague that it could not be quantitatively tested or discussed and then offering quantitative calculations intended to disprove it—and making errors even in those calculations [173, 391, 436];

By exhibiting jealousy of Velikovsky: *Worlds in Collision* came out "at a time when the work of scholars cannot be published for lack of funds" [287];

By making such sloppy mistakes as misspelling names [236, 293, 391], misquoting Velikovsky [236], criticizing points on which established authorities agreed with Velikovsky [72, 173], and misrepresenting Velikovsky's argument [73, 185, 210].

And what can one say about an individual [155] who doesn't believe Velikovsky but does believe in the canals of Mars?

So the criticisms of Velikovsky are replete with committed blunders. I would add that the critics were also guilty of important errors of omission: I have shown that *Cosmos without Gravitation* can readily serve as evidence that Velikovsky was ignorant of the fundamentals of physical science; could not this demonstration have been given in 1950? It appears that only two reviewers [214, 288] referred to that monograph, and then only in passing. But the greatest error of omission of which Velikovsky's critics were guilty was the failure to challenge the claim, so frequently made [80, 189, 389] (Chapter 1, *Previews and Summaries;* Chapter 2, *Reviews*), that Velikovsky's work displayed a high order of originality. Thus the *American Behavioral Scientist* [71] referred to "the new historical and cosmological concepts of Dr. Immanuel Velikovsky." How new, in fact, were these concepts?

Velikovsky's Originality

The salient features of Velikovsky's method and conclusions are these:

1. The Bible and other ancient records contain much factual material about actual physical events.

2. Myths and folklore are built around descriptions of striking actual events that can be discerned by proper analysis.

3. Past catastrophes were so threatening to mankind that overt reminders of them are suppressed, though they are described in more or less veiled form in common legends and traditions. The refusal to see descriptions of these actual catastrophes in the ancient writings is a result of psychological suppression, of collective amnesia.

4. The catastrophes resulted from near approaches of heavenly bodies to the earth.

5. Legends of gods fighting and courting reflect actual encounters between heavenly bodies.

6. The major catastrophes resulted from near approaches of a comet, which produced world conflagration, a long period of darkness, floods, and hurricanes; the duration of the year was altered (it was at one time exactly 360 days); the earth's axis tilted, or its rotation slowed, during Joshua's miracle; electromagnetic forces were important.

7. The comet settled into orbit around the sun as the planet Venus.

8. There is a continuing terror of comets among mankind as a result of these unconsciously remembered events.

9. The catastrophes can be used to synchronize the histories of different cultures. The conventional chronology of Egypt is wrong (too long); the Greek Dark Ages are an invention.

Now, it turns out that *only one of these assertions is new,* the identifying of the comet with what is now Venus. Clear evidence of this lack of originality is present in the literature on the Velikovsky affair itself (see below). But Gardner [112, 113] was the only early critic to stress that the claims of Velikovsky's originality were spurious. More recently, Carl

Sagan [353, 356–59, 361] has stated (but not illustrated) the point.

"[That] stories reported as direct observation in the ancient chronicles are strictly true" was quite a common assumption in earlier times, and "attempts to find some physical explanation, however bizarre" were also made. "The last important scientific work that used Velikovsky's method . . . [was] Thomas Burnet's *Sacred Theory of the Earth,* first published in the 1680s" [127]—cf. number 1 above.

". . . other searchers for meaning in ancient sources . . . the epic-minded Ignatius Donnelly, the serio-humorous skeptic Charles Hoy Fort, the deterministic Isaac Newton Vail, the cataclysmic George McCready Price, the weird ice-age theorist Hans Hoerbiger, and many others . . . have pointed to 'stones, bones, and runes' as evidences of catastrophic upheavals in man's past" [166]—cf. number 1 above.

Giambattista Vico in *The New Science* (1725) "insisted that . . . the myths of earlier times ought to be taken seriously as themselves accounts of the actual history of those times . . . ," and "further proposed that . . . when rationality became ascendant, a collective amnesia took place; men denied that their own historical origins were contained in those myths . . ." [285]—cf. numbers 2, 3.

". . . Bruno propounds an interpretation of ancient astromythology that is similar to that followed by Velikovsky" [379]—cf. numbers 2, 5.

"Kugler wanted to communicate to the public . . . '. . . that ancient traditions, even when . . . dressed as myth and saga, cannot be dismissed lightly as fantastic, or worse, meaningless fabrications . . .' "; "Kugler arrives at the conclusion that the saga of Phaethon has as its historical core the appearance of a comet that was followed by a partial world fire and a flood. . . . this interpretation has been already offered several times in antiquity. . . . The first references to it are in Plato and Aristotle . . ." [379]—cf. numbers 2, 6.

"Nicolas of Cusa (1401–1464), in his *De docta ignorantia* . . . claimed that heavenly motions do not have stability as an inherent quality, and formulated the hypothesis that some statements of ancient writers may be explained by their having seen a sky different from what was seen in his time . . ." [378]—cf. numbers 1, 4.

". . . Nicolas-Antoine Boulanger (1722–1759) . . . analyzed the cosmogonies and mythologies of several farspread peoples of the Earth, such as Germans, Greeks, Jews, Arabs, Hindus, Chinese, Japanese, Peruvians, Mexicans, and Caribs, concluding that rites, ceremonials, and myths reflect the fact that the human race was subjected to a series of cosmic convulsions for which he also considered the geological and paleontological evidence. He argued that these catastrophes shaped the human mind, causing among other things a deepseated psychological trauma . . ." [378]—cf. numbers 1, 2, 3, 4.

"Laplace stressed that the human race is beset by a great fear that a comet may upset the Earth. . . . He proceeded to describe the possible effects of a collision with a comet, painting a picture that is in close agreement with that outlined by Velikovsky. . . . Laplace also wondered whether heavenly bodies might not be affected by forces other than gravitation, such as electric and magnetic . . ." [378]—cf. numbers 6, 8.

". . . Newton . . . argued that accepted chronology must be lowered and anticipated the conclusions reached by Velikovsky in *Ages in Chaos.* Like Velikovsky, he claimed that Greek chronology must be shortened by four hundred years, eliminating what today are called the Dark Ages of Greece. Like Velikovsky, he claimed that some dynasties of Egypt have been duplicated in chronological schemes . . ." [378]—cf. number 9.

"Like many of his contemporaries, Newton was concerned to reconcile the chronologies of ancient Egypt, Babylonia, Assyria and Greece with Biblical chronology . . ."; like Velikovsky, he recognized no Greek Dark Ages and dated the Hyksos invasion of Egypt after the Exodus [110]—cf. numbers 1, 9.

William Whiston, in *A New Theory of the Earth* [461], considered that a comet could become a planet. He drew from the Bible and from legends of many cultures the inference that the close approach of a comet to the earth had caused the Deluge and changed the earth's orbit and rotation. He postulated electric and magnetic effects during the encounter [113:32–35]. Whiston dated the Exodus to 1489 or 1490 B.C. [461:142]. His evidence that the year was at one time 360 days in duration was similar to Velikovsky's, for example that 150 days elapsed from the 17th day of the second month to the 17th of the seventh month [461:162]. The similarity between the work of Velikovsky and that of Whiston was noted by Gardner [113:32–35], Moore [256:58], and Stecchini [378].

The idea that earth's history has been a catastrophic one was generally accepted until the nineteenth century, when it was replaced by the uniformitarian, or gradualist, ideas to which Velikovsky frequently refers as the theories of Lyell and Darwin. Comets have often been suggested as causes of terrestrial cataclysms, and the idea was quite common even in the late nineteenth century [323]. Thus Alexander von Humboldt, looking back, felt that "geology before M. W[erner's] time . . . was a succession of daydreams on comets which caused deluges and conflagrations of the surface of the globe" [54]. The close approach of a comet to the earth has remained a frequent theme in science fiction. Aldiss mentions, for example: *Hector Servadac* or *Off on a Comet* by Jules Verne—"A comet has grazed the Earth . . ." [6:97]; *Olga Romanoff* by G. C. G. Jones—". . . a passing comet burns almost everyone to a cinder . . ." [6:142]; *The Day of Uniting* by Edgar Wallace—". . . the threat of a large comet which, in its close approach to Earth, causes humanity to unite . . ." [6:153]; and the capture of a comet by the sun is described in *The Great Stone of Sardis* by Frank Stockton [6:143].

In 1883 Ignatius Donnelly published *Ragnarok* [91], to which *Worlds in Collision* bears many striking similarities—in mode of thought, in the sources and their interpretation, and even in the impact of the book on the public and on the scientific community. Like Velikovsky, Donnelly surveyed existing scientific knowledge, pointing to gaps in it and evidence for past catastrophic events—for instance, quoting the same passage from Cuvier as does Velikovsky concerning extinction of the mammoths (*Ragnarok,* p. 47, and *Worlds in Collision,* p. 25). Like Velikovsky, Donnelly described what effects the approach of a comet to earth would produce: a change in inclination of the earth's axis; great cracks in the surface; inconceivably powerful winds; vast heat; the fall of debris producing darkness, electrical effects, subsequent rain and cold [91:94–111]. Like Velikovsky, Donnelly claimed to have found descriptions of such events in the legends of Asia (Hindu, Chinese, Burman, Polynesian), Europe (Druid, Greek, German, Spanish, Eskimo, Scandinavian), Egypt, the Middle East (Persian, Phoenician, Babylonian, Syrian, Hebrew, Assyrian, Moslem), Africa, South America, and Central and North America. The similarities of detail are striking:

"Acted upon by magnetism or electricity . . . in the tail of the comet" [91:45], the latter appeared in the sky and was described for posterity as a bear, a snake, "the dragon, the serpent, the wolf, the dog, the Evil One" [91:132, 429]; "all the winged beasts breathing fire are simply a recollection of the comet" [91:119]—cf. pp. 78 and 305 in *Worlds in Collision.*

"This is the event that lies, with mighty meanings, at the base of all our theologies" [91:250]—cf. p. 380 in *Worlds in Collision.*

". . . there is an untaught but universal feeling which makes all mankind regard comets with fear and trembling . . ." [91:424, 430]—as in *Worlds in Collision,* p. 303.

Donnelly talks of iron-stained red clay cast down from the comet [91:149]—compare *Worlds in Collision,* p. 48 (rusty pigment, ferruginous, red).

The presence of carbon in comets was known from spectra, and Donnelly suggested that carbureted hydrogen, the "deadly fire-damp of the miners" (i.e. methane), might have been the poisonous gas that reached earth from the comet [91:104–6, 134, 147]. Velikovsky also talked (mistakenly) of methane in this respect as a poisonous gas [408:366–71].

Many references are common to both books—Hesiod, the Norse Fenris-Wolf, the myth of Phaethon from Plato and Ovid, the Popol Vuh, the Zend-Avesta, and so on. Both reference, for example, *The Great Ice Age* by Geikie and *The Native Races of America* by Bancroft.

Donnelly [91:264] wrote: "It is, at least a curious fact that in Joshua (chapter x) the standing-still of the sun was accompanied by a fall of stones from heaven. . . ." This connection was made much of by Velikovsky as the starting-point [408:40–44] of his discussion.

Donnelly [91:85] wrote that "Lexell's comet . . . first came within the influence of that planet [Jupiter] in 1767; it lost its original orbit, and went bobbing around Jupiter until 1779, when it became entangled with Jupiter's moons, and then it lost its orbit again, and was whisked off into infinite space. . . ." He quoted from the *Edinburgh Review* (October 1874, p. 205) that "in the years 1767 and 1779 Lexell's comet passed through the midst of Jupiter's satellites, and became entangled temporarily among them. . . ." Velikovsky [408:78] wrote of "Lexell's comet, which in 1767 was captured by Jupiter and its moons. Not until 1779 did it free itself from this entanglement. . . ." He gave no reference for this information, but the uncommon usage of "entangle" suggests that it was the same as Donnelly's.

As noted by Gardner [112], "Velikovsky's brief footnote reference to . . . [Donnelly's *Ragnarok*] does not indicate how remarkably similar it is to his own work. . . ." Velikovsky's footnote reads:

> While investigating whether an encounter between the earth and a comet had been the subject of a previous discussion, I found that W. Whiston . . . tried to prove that the comet of 1680, to which he (erroneously) ascribed a period of 575½ years, caused the biblical Deluge on an early encounter.
>
> G. Cuvier, who was unable to offer his own explanation of the causes of great cataclysms, refers to the theory of Whiston in the following terms: "Whiston fancied that the earth was created from the atmosphere of one comet, and that it was deluged by the tail of another. The heat which remained from

> its first origin, in his opinion, excited the whole antediluvian population, men and animals, to sin, for which they were all drowned in the deluge, excepting the fish, whose passions were apparently less violent."
>
> I. Donnelly, author, reformer, and member of the United States House of Representatives, tried in his book *Ragnarok* (1883) to explain the presence of till and gravel on the rock substratum in America and Europe by hypothesizing an encounter with a comet, which rained till on the terrestrial hemisphere facing it at that moment. He placed the event in an indefinite period, but at a time when man already populated the earth. Donnelly did not show any awareness that Whiston was his predecessor. His assumption that there is till only in one half of the earth is arbitrary and wrong. [408:40–44]

The only references to Donnelly and Whiston in the index of *Worlds in Collision* are to this footnote and to one on p. 330 that mentions Whiston's belief in a 360-day antediluvian year. Thus the reader of *Worlds in Collision* is given the impression that these other authors merely speculated—on unspecified grounds—concerning an encounter of the earth with a comet, and that they were both wrong on major points, thus vitiating their work; the great and many similarities to Velikovsky's approach, the similar use of the same legends and references, are not acknowledged. Velikovsky's debt to these authors is far greater than these references to them in *Worlds in Collision* indicate. It is clear that Velikovsky had read *Ragnarok* closely; for example, he gives it as authority for material from the annals of Wong-shi-Shing [408:118, 131] and for Aztec [408:118] and Oraibi [408:122] folklore.

As early as 1948, before *Worlds in Collision* had appeared, Velikovsky disclaimed any great debt to his precursors [451]: "There are three more authors [Whiston, Cuvier, Donnelly] whom I feel obliged to mention, although I did not find in them more than one or another quotation which I could trace and re-employ, since my theory was ready when I came across their books, and it went far beyond the ideas of these authors." And he seemed to say that it is not uncommon to fail to acknowledge those who earlier put forth similar views:

"Thus, the authors of basic discoveries of the laws of nature borrowed from the ancients or rediscovered the truths known to the ancients, but kept the ancients' names out of their discourses."

Regardless of whether Velikovsky copied much or any of his material, it is clear that there is not much basis for lauding his originality in all this. He followed others in finding fact (even the *same* fact) in myths; comets as dragons menacing the earth with fire, flood, and stones; collective amnesia of the catastrophes; the details of the catastrophes; the origin of religions there. All the generalizations and most of the particulars that, many commentators had said, originated with Velikovsky can be found in earlier works, particularly[1] in *Ragnarok.* And *Ragnarok* had even been received much as was *Worlds in Collision:* "it was an immediate success with the general public. Scientists ignored it, but the reviews in popular journals were favorable" [113:32–35]. A review printed at the back of the book [104] is strikingly reminiscent of what was later to be written about *Worlds in Collision:*

> It is not an exaggeration to say that "Ragnarok" is one of the most original and striking productions of recent times. . . . The work consists of a chain of arguments and facts to prove a series of extraordinary theories . . . that the legends of all the races of the world preserve references to and descriptions of this catastrophe. . . . These startling ideas are supported by an array of scientific facts, and by legends drawn from all ages and all regions of the earth.
>
> The book will, of course, encounter the hostility of those who have devoted their lives to proving that all the gigantic phenomena of the Drift were due to the action of ice; but, whether Mr. Donnelly has or has not established his theory of the contact of the earth with a comet, he certainly makes many strong points against the ice-theory.

1. The Macmillan files contain a letter from O'Neill, dated March 1, 1947, which is intriguing in this connection. O'Neill had just come across *Ragnarok* and now wrote to Macmillan in amazement that Velikovsky's manuscript had no "single word of reference by Velikovsky to Donnelly's work . . . that Velikovsky did not make extensive mention of Donnelly."

> A notable part of the book is its relation to the Bible. It claims that . . . the book of Job is a description of the same great event. . . . the narrative of the book contains a very accurate description of the events of the Drift Age and the comet itself.
>
> But no mere summary can do justice to this remarkable book. Whether readers believe in it or not, they will find it exceedingly entertaining. . . .

Tit for Tat

So far, I have emphasized how ineffective Velikovsky's critics have been. I pointed out that the *ad hominem* attacks on Velikovsky actually served to bring others to his support; I did not stress that *ad hominem* attacks should not be countenanced. In part, my emphasis was determined by one purpose of this book—to make plain what scientists must do, and what they must not do, if they are to be effective in public controversies. But in some part also, my emphasis stems from the fact that I wrote this book after concluding that Velikovsky's cosmic scenario is too implausible for me to take seriously: I tend to allow more latitude to those who seem to me right than to those who seem to me wrong, I am more critical of the improprieties of those who seem to me wrong than of those who seem to me right (see Chapter 15, *Ethics in Science*). If, however, I make the effort to lay aside altogether the possible merits of Velikovsky's work, then the critics of Velikovsky come into focus in an even harsher light. I find in the record that the critics were themselves guilty of many of the things for which they castigated Velikovsky and his supporters.

Velikovsky's ignorance of science was criticized, ignorance of the subjects he discussed. Yet some of the critics did not hesitate to discuss *Worlds in Collision* without having read the book, rather clearly in ignorance of the subject they talked about. And critics of subjects on the fringes of science continue to be guilty of this (for example, regarding astrology [206, 458, 459]). Carl Sagan, for instance, evidences few inhibitions in making sweeping, simplistic generalizations in areas where he is hardly expert. Historians, philosophers, and sociologists of science could well take exception to his statements [356] that "science is . . . self-correcting . . . appeals to authority are impermissible . . . accepted theories and hypotheses have been entirely overthrown. . . . This self-questioning and error-correcting aspect of the scientific method is its most striking property, and sets it off from many other areas of human endeavor, such as politics and theology." These statements all need qualification if they are to be taken seriously; dare I suggest, perhaps as much qualification as Velikovsky's analogy between atom and solar system, for example? And Sagan quotes "a distinguished professor of Semitics at a leading university" about myths and legends, but then strikes out on his own: "let me not be swayed by the opinions of others"—not even where the others are expert and Sagan is not. This is precisely the starting *ab initio* in established disciplines for which I have criticized Velikovsky.

Quite inexcusable were the innumerable occasions on which critics of Velikovsky misquoted or misrepresented him (see above and Chapter 14, *Word Games* and *The Numbers Racket*). It seems incredible that salient points of Velikovsky's scenario should not be known to those who comment on those specific points, yet that continues to be the case. For example, Velikovsky gave no precise date for the fission or expulsion of comet-Venus from Jupiter, but made clear that it was long before the postulated encounter of Venus with earth about 1500 B.C. Yet critics continue to confuse the latter date with the former [2, 65], or even to say that Venus was ejected from Jupiter "2500 years ago" [19], a millennium *after* its postulated encounter with earth. In another example the whole controversy about Velikovsky was said to be "dormant during the hectic days of the '60's" [19], and "Velikovsky's star has been in eclipse since . . . 1950" [328], when the 1960s were actually years of intense public debate (Chapter 4).

Velikovsky was often criticized for being too simplistic. So are the statements of Sagan quoted above, and his attempted calculation of the probability of Velikovsky's scenario, for example. Those calculations have been properly criticized a number of times [100, 192, 193, 259, 332], by such qualified and impartial individuals as Robert Jastrow [160]. The estimating of the probability of a very unlikely event is a most fallible pursuit (see pp. 313–14); the translating of the calculated low probability into a judgment that the event will not occur, or could not have, is completely fallacious. If an event has a probability greater than zero, then it can happen. Sagan's penchant for simplicism in these calculations is demonstrated further by the fact that he applied the same equation to estimating the size of the population of Loch Ness monsters, treating as a "collision" the sighting of a monster by an observer, with the "collision radius" corresponding to the observer's field of view [354].

Velikovsky was criticized for being dogmatic, for claiming to be never wrong. Chapter 8 offers many examples of science writers and scientists who dogmatically declare something to be pseudo-science, and who claim never to be wrong in their declaration (see also Chapter 1, *First Criticisms,* and Chapter 2, *Reviews*). Velikovsky's method was criticized as being "apologetic": searching out evidence to support his prior belief. It is plain enough that many of Velikovsky's critics had found him wrong before finding the proof of that—certainly those who had not read his work—and, I imagine, Sagan did not calculate his probabilities before judging Velikovsky wrong.

Velikovsky's interpretations are farfetched, implausible, invalid. What about the inference [274] that our high standard of living proves our science to be right? What about the interpretation of Bode's law (Chapter 5, *Scholarly Support*) as an argument against Velikovsky's scenario?

Velikovsky was taken to task for not properly answering the criticisms of his ideas. But Menzel (Chapter 4, *Something Amiss in Science*) failed to correct some erroneous calculations that he wished to use against Velikovsky. And Sagan has continued to publish his critique in slightly different variants [353, 356–59, 361], none of them taking cognizance of numerous published objections [100, 132, 160, 192, 193, 259, 332]. At least one of the changes he has made seems almost surreptitious. In his early version he says "I do not believe that all of the concordances which Velikovsky produces can be explained away in this manner" [356]; lately he says "I believe that *all* of the concordances Velikovsky produces can be explained away in this manner" [358, 359]

That last example could also be mentioned to those who accused Velikovsky of lack of self-consistency (Chapter 2, *Reviews*), and to the critic who wrote, "Anyone with even a modest training in astronomy or physics would recognize the theory [Velikovsky's] as absurd," only to continue a few paragraphs later, "he [Sagan] nevertheless takes Velikovsky seriously" [2].

I myself have noted discrepancies between Velikovsky's self-image and the actualities (Chapter 9, *Self-Image*). What about the scientists who claimed for science, and thereby for themselves, open-mindedness and objectivity? The resistance of science to very novel ideas is an established phenomenon, and Velikovsky's supporters were more nearly correct in recognizing that (Chapter 15, *Resistance by Scientists to New Discovery*).

I have referred to wishful thinking on the part of Velikovskians (Chapter 12); that has been displayed also by the critics, for example, "Velikovsky is flatly and totally disproven" [19].

In Chapter 14 I illustrate how tendentious was much of the arguing in the Velikovsky affair and find no difficulty in drawing examples from both critics and supporters. But in some types of impropriety I find the one side guilty and the other innocent. Critics of Velikovsky boycotted a publishing house in the attempt to suppress a book: the Velikovskians have not done that. Critics of Velikovsky were instrumental in the loss of livelihoods by Atwater and Putnam: the Velikovskians have not so harmed anyone. Journals of science have refused to publish articles, letters, and advertisements from supporters of Velikovsky: *Pensée, Kronos,* and *S.I.S. Review* have given space to critics of Velikovsky and have even solicited contributions from them [302, 348, 386, 387].

14

Means of Persuasion

> Christians have burnt each other, quite persuaded
> That all the Apostles would have done as they did.
> —Lord Byron

Shades of Meaning

Commonly in arguments one wants to win: not infrequently, to win irrespective of the means that might become necessary (limits being set variously by different people; knives and guns are still regarded as the natural means in some places, including some parts of the United States). In the Velikovsky affair a wide variety of verbal weaponry was used: innuendo, implication, misleading statement, misquotation or misinterpretation or misrepresentation of the opponent; argument by analogy, *non sequitur, ex cathedra,* and *ad hominem;* wishful thinking; sarcasm and ridicule. Perhaps the greatest difficulty in attempting to make sense of such a controversy is that so many important influences are exerted by innuendo, shades of meaning, statements taken out of context, apparently weighty yet actually meaningless facts, and such influences are not always readily detectable. For instance, I doubt that many people realized at the time what effect (and how potent a one) Velikovsky achieved simply by presenting his work as "heretical" [408:preface]. By this self-definition, opposition to his views was naturally to be expected, so when it came, Velikovsky was able—quite logically, granted his premise—to maintain that, just as he had forecast, the dogmatic establishment wished to suppress him. Thus Velikovsky turned the criticisms of his work into an argument in his favor; the reviewers who attempted to point to various errors had already been branded dogmatists.

In hindsight one sees that, to be effective, Velikovsky's critics would have had to destroy his self-proclaimed status of arch-heretic: to show that he was not unique, not even unusual, not particularly original, that his ideas lacked significance as well as being erroneous. The status of heretic inevitably implies high significance and makes plausible the comparisons with earlier heretics who were later vindicated. Unless Velikovsky could be shown to be without significance, he could not effectively be shown to be wrong. As it was, many were no doubt influenced by the repeated comparisons of Velikovsky with those previous heretics. Many did not see that this was a shifting of the burden of proof from Velikovsky—where it belonged—to his critics.

Similar examples of implication abound throughout the Velikovsky affair. *Caveat emptor* is perhaps the most important rule if one wants to be able truly to make up one's own mind: beware what one buys, when one reads too casually, too superficially, to detect the all-important subtleties of implication. Here is an example, a passage I had to read several times before I saw what bothered me about it. The author is talking about fringe ideas, crackpot and crank ideas: "Not all the ideas discussed in the following chapters are equally foolish. Some, like extrasensory perception and the Loch Ness monster, may ultimately turn out not to be foolish at all; indeed, it is the skeptics who may look silly in the end. . . . There is much that is unknown, and we must keep an open mind toward all novel ideas. But an open mind does not mean a suspension of disbelief, for scientific proof is not poetry. We must be able to evaluate evidence and if it is found wanting, to discard it" [51:5–6]. That surely seems harmless enough; the author seems quite careful to admit fallibility. But what is that word "foolish" doing there? How about "skeptic"? Does the last sentence really warrant the emphatic and absolute "must"?

I could agree with the first sentence if "equally foolish" were replaced by "necessarily wrong," and with the second if "foolish" were replaced by "wrong." "Foolish" is an expression of value, of opinion; as it stands, the second sentence is not really the admission of fallibility that it purports to be but, rather, reinforces the author's contention that all these ideas are indeed foolish now, in the light of what is known at the time he writes—that only fools believe them, that one is a fool if one believes. Yet the author admits that some of these ideas may turn out to be right after all. He says, in effect, that he himself knows what is compelling evidence and what is not, and that it is foolish to entertain ideas that have not yet been certainly established but that may turn out to be valid.

The *Oxford English Dictionary* defines skeptic as "one who maintains a doubting attitude with reference to some particular question or statement. Also, one who is habitually inclined rather to doubt than to believe any assertion or apparent fact that comes before him; a person of sceptical temper"; "occasionally used with reference to the etymological sense: A seeker after truth; an inquirer who has not yet arrived at definite convictions." Now, how could one who is so judicious be seen as "silly"? Only if the evidence had been completely, absolutely, objectively compelling and yet he remained skeptical. But by the author's own contention that is not the case here: it would be "foolish" to believe now. So the author is labeling as "foolish" and "silly," first, individuals who happen to be wrong (which makes them neither foolish nor silly) and, second, some who by his own reasoning cannot properly be so labeled. But I had to think carefully and clearly before I recognized that the author was asserting evaluations with which I do not agree. So it behooves us to ponder such words as "foolish" and "silly" as we read them, lest we unknowingly imbibe the unsupported opinions of someone else.

Again, in the last sentence of that quoted paragraph is hidden a mass of assumptions: that objective evaluation of evidence is possible; that only ideas supported by compelling evidence ought to be entertained seriously; that scientific proof is absolute proof. The repeated "must" implies dire consequences for those who do not follow the author's precepts. Now I happen to agree with some of the author's assumptions but the point is that, unless one is careful, one buys the assumptions without even realizing it.

There are many similar pitfalls in the literature of controversy. Most of the time we are presented with material that is slanted—either deliberately or unintentionally, often a bit of both—in order to convey an interpretation in the guise of facts, information. Not so long ago *Time* was justly famous for its practice of wording reports in an insidious manner; that style is nicely exemplified by the treatment of Velikovsky in 1950 by that magazine [400]. I cannot resist making the suggestion that we coin a term for this type of activity—*newsmanship:* the art of conveying a false impression without actually lying. The Velikovsky affair is pervaded by newsmanship, some of it of the highest caliber.

Word Games

Argumentation in the Velikovsky affair ran the gamut from such relatively subtle implications as those discussed above all the way to promulgation of plain untruths.

Looking back, it seems clear that Velikovsky suborned the truth through such devices as claiming, in the face of the facts, to stand as a noble heretic, as above the verbal fray, as unconcerned about being given credit and priority for his ideas; misleadingly invoking the names of historical and contemporary celebrities; harping on dogmatism and psychology as reasons for his rejection by scientists; demanding to be taken all or nothing, as "fiction or nonfiction." Consistently, Velikovsky implied that his ideas must have merit because of the resistance they met, a *non sequitur* if ever there was one. He shifted ground adroitly as needed, maintaining that his critics were attacking some "theory of catastrophism" rather than his own specific works [4, 103]. Velikovsky argued by analogy, *ex cathedra, ad hominem,* and through rather plain untruths: his quotations out of context from Humphreys and Hulburt (pp. 110–11), Jeans (pp. 116–17), McCrea (p. 130). Into this last category fall his statements that he had left no criticism unanswered [410:foreword]; that no changes in his books are called for by the advance in scientific knowledge [445]; that the second law of thermodynamics makes impossible a "greenhouse" effect as an explanation for the temperature of Venus [445].

Nor could one readily detect the truth by listening to Velikovsky's critics. A common piece of newsmanship here was the explicitly stated difficulty of discerning a crank, phrased in such a way as to induce the reader to believe that in this particular case the identification was certain (Chapter 8, *Label with Care*). And, again, there were what I submit to be outright lies: that Velikovsky called a planet a comet [287]; regarding Velikovsky's suggestions about Joshua's miracle [73]; Shapley's denial that there had been any sort of campaign waged against Macmillan [90, 272]; that the latter had intended *Worlds in Collision* as a textbook [17]; that Velikovsky had suggested global cataclysms as occurring without anyone's nose even bleeding [279]; and so on and on.

Isaac Asimov [17] misreported several aspects of the affair, and his mode of emphatic writing with sweeping generalizations gives the reader scant opportunity to choose his own views from the facts at hand: "if anyone reads *Worlds in Collision* and thinks for one moment that there is something to it, he reveals himself to be a scientific illiterate" [16]—in the company, no doubt, of assorted scientifically illiterate professional scientists who have given more than a moment's thought (for example, Bass [24, 25], Burgstahler [42–45], Michelson [248–50], Miller [251], Ransom [330–34]). In *Nature* [340] one could read that "there seems little danger that any real scientist or scholar would be tempted to take Velikovsky's ideas seriously . . ."—a gross insult to those real scientists and scholars; that "best-selling books are usually fiction"—no doubt with reference to such works as the Bible and Darwin's *Origin of Species;* and that the atmospheric clouds of Venus are "now [1974] known . . . to consist of sulphuric acid"—something that had indeed been suggested but was far from known.

Ley [219] informed readers that Velikovsky was wrong, supporting his statement by merely asserting an analogy with Hoerbiger and saying that "a mathematical analysis of Velikovsky's statements shows the complete impossibility of every one of his assertions"—simply unwarranted dogmatism spiced with a bit of the numbers racket, rhetorical invoking of mathematics. Patrick Moore, who has written excellent popular works on astronomy, gave a quite unfair view of Velikovsky when he wrote that "to Dr. Velikovsky, astronomical time is to be measured in thousands of years rather than in millions" and that "the fossil record of the Earth goes back a great deal further than Biblical times" [256:59, 61]—as though Velikovsky had ever questioned that. In an editorial in a science fiction magazine [13] Velikovsky's claim of carbohydrates in the atmosphere of Venus was criticized; Velikovsky never made such a claim—he referred to hydrocarbons, quite different substances.

Charles Fair [106:171] made the extraordinary statement that ". . . Velikovsky does not tell us where he got the idea that Venus originated as a comet. . . ." Admittedly, that is true in a strictly literal sense—how many of us know where or how we get an idea? But no scientist is ever required to relate how he came by an idea: it is the discussing and buttressing of the idea with evidence that is of concern. Velikovsky certainly wrote voluminously about references that he believed to support his idea: pp. 153–203 in *Worlds in Collision,* for example. What more could one ask?

Velikovsky's supporters, of course, were as guilty as Velikovsky himself and his critics. An added difficulty here was that one could not readily distinguish who was a supporter, and who a relatively unprejudiced observer, at any given time. So in 1963 there would have been no reason to doubt that the *American Behavioral Scientist* represented people who had a "public" interest in the controversy, yet with the perspective of hindsight one now recognizes de Grazia, Juergens, and Stecchini as advocates rather than commentators or observers. The tendentious nature of the discussions in the *American Behavioral Scientist* and *The Velikovsky Affair* is illustrated by the comment of a reader that "only after The Velikovsky Affair was published . . . did I learn that because this man had the impudence to trespass into other disciplines . . . and because the scope of his work was so breathtaking, certain senior academics in the United States became transformed into what can only be described as megalomaniacs" [123].

Much was made of the fact that Velikovsky's critics had not read his work before criticizing it [172, 378]. There is certainly some evidence to that effect, in the cases of one or two individuals who admitted as much, and perhaps of those who criticized solely on the basis of previews and summaries before *Worlds in Collision* was available to them. On the other hand, the situation is really not as clear-cut as the Velikovskians would have it. Shapley was accused of not looking at Velikovsky's evidence, but Juergens's own account [172] makes clear that Velikovsky had met Shapley years earlier and had discussed his ideas with him, in sufficient detail to ask him to carry out observational astronomical tests. Velikovsky had also sent copies of *Cosmos without Gravitation* to various astronomers and academic institutions.[1] Perhaps Shapley had not read *Worlds in Collision,* but he was certainly acquainted with the gist of Velikovsky's ideas. Larrabee [210] accused Payne-Gaposchkin of not being familiar with Velikovsky's notions about gravity, yet she had indeed read *Cosmos without Gravitation* [288]. So, with hindsight, we recognize that the refrain of "criticism before reading" is not quite the whole truth.

Larrabee [211] led the unwary astray by stating that a hot Venus with atmospheric hydrocarbons was a necessary consequence of Velikovsky's theory (whereas those are not necessary consequences objectively demonstrated but merely

1. Presumably including Harvard: Payne-Gaposchkin [288, 290] had read the book. The copy I obtained from the library at Johns Hopkins University was inscribed as a gift from the author.

personal intuitions of Velikovsky); and that the successful advance claims were contrary to prevailing views, implying a contradiction with the body of existing theory, whereas in fact these were simply minor points of detail which can without difficulty be explained in conventional ways. (Note that giving an explanation for virtually anything is usually quite easy; the difficult thing is proving the explanation to be correct. My scientific mentor was wont to say that he did not much care what results my experiments would produce—that he could just as easily and happily explain a negative result as a positive one. And indeed he could, quite plausibly.)

In 1976 a review of *Velikovsky Reconsidered* in *New Society* [374] stated that "for almost 30 years now, Velikovsky has been persecuted with a McCarthy-like intensity by a community which prides itself on objective consideration of both sides of an argument. . . ." But could a reader of that periodical realize that the "persecution" consisted in the main of not accepting the views of someone who failed to present them in the accepted scholarly manner to which all other scholars submit as a matter of course?

According to *Philadelphia Magazine* [233] in 1968, "Velikovsky was . . . relegated [by the scientific establishment] to a silent limbo in which he became a non-person. This punishment . . . persisted for more than a decade." That sounds like an academic sent into the wilderness, perhaps through losing an academic post. But Velikovsky had never held an academic post and had never been anything but a nonperson in the disciplines of the sciences and history.

Again, "Unmindful that their own fraternity had denied him access to scholarly journals, the scientists accused Velikovsky of publicity-mongering." But Velikovsky had no access to scholarly journals simply because he did not write scholarly papers. By his own admission, he refused to revise his work as suggested by referees or even to argue against the case made by the referees: "My answer was submitted to *Science* editors, was returned for rewriting after one or two reviewers took issue with my statement that the lower atmosphere of Venus is oxidizing. I had an easy answer to make. . . . But I grew tired of the prospect of negotiating and rewriting . . ." [425]. So Velikovsky's exclusion from scholarly journals is not quite the indictment of those journals that the Velikovskians have tried to make it. Here, as almost everywhere else, there is something to be said on both sides.

Velikovsky rightly acknowledged [445] that "what I offered is primarily a reconstruction of events in the historical past. Thus, I did not set out to confront the existing views with a theory or hypothesis and develop it into a competing system. My work is first a reconstruction, not a theory. . . ." But everyone conspired to obfuscate that plain and important fact: Velikovsky himself by speaking of his theory whose "consequences . . . affected almost all natural sciences and many social disciplines" [433] and of "the theory of cosmic catastrophism" [408:384], and his critics [52], his followers [73], and people acting as commentators [92, 209, 362] who referred to these "theories."

That Velikovsky spent ten years in research I have shown (Chapter 9, *Never Wrong*) to be rather misleading. This implied untruth was promulgated not only by Velikovsky but also by his proponents and by many apparently disengaged observers [18, 92, 217, 218, 284].

As regards Venus, Velikovsky's advance claims in *Worlds in Collision* did *not* include a statement that Venus rotates in retrograde fashion; he quoted extant estimates for the rate of rotation without giving any inkling of a suspicion that the sense of rotation might be unusual. Yet the casual observer has been exposed to the plain untruth that Velikovsky had specifically claimed retrograde motion for Venus—an untruth indulged in by Velikovsky himself [424, 435]; by Velikovskians [171, 468]; by a reviewer of *Velikovsky Reconsidered* [235]; by a reporter for *Industrial Research* [154]; by the otherwise quite reliable MacKie [230] and Michelson [250]; by a well-known, popular writer on astronomy, Patrick Moore [256:62]; in a compendium of anomalies [148:31];

and even by that harsh critic of Velikovsky, Carl Sagan, who failed to expose the lie when he had the opportunity to do so [122].

The point is that untruths of varying degrees are promulgated (not necessarily deliberately, of course) by all parties in these public controversies. Once even a small inaccuracy is left unexposed, its repetition tends to accrue further inaccuracies, and the final effect can be totally misleading to one who is not completely familiar with the issues and the literature. I believe that when Venus's retrograde motion was first discovered, Velikovsky probably intended to say no more than that this was consistent with his view that Venus was a recent member of the system of planets. In repetition in shorter form, this became the very different statement that he had made a much more specific advance claim than he actually had, and the force of that on the casual reader must be considerable.

Again, Latham [213:71–77] wrote from the perspective of a couple of decades about the beginnings of the Velikovsky affair, and rightly said that Macmillan had never advertised *Worlds in Collision* as a scientific text. But he went a bit too far when he said that no scientific claims were ever made for it: the book was, after all, listed "in the Macmillan spring catalogue under the heading 'Science.' . . ." [221]. That inaccuracy of Latham's could give the reader a false impression about the attitudes of some, at least, of the scientists who plagued Macmillan with their protests.[2]

Jueneman [166] remarked that Velikovsky had been accused of publishing simply in order to make money, yet I have been unable to find that accusation in the published material. It is perhaps the only unfair criticism *not* made of Velikovsky.

Doubleday quoted out of context to make unfavorable reviews appear favorable. Thus, from the *Times Literary Sup-*

2. The details of this issue have been uncovered by Ellenberger [99B]. His account confirms that both sides overplayed their hands in this case.

plement: "Dr. Velikovsky is a man of much learning and he has mastered the commentaries upon the text and all the books on cosmology as well . . ." [92N]; the publisher omitted the sarcasms that follow immediately upon that sentence, as well as "it is idle to apply the standards of critical scholarship to this work" and "No doubt the lunatic fringe and the fundamentalists will respond . . ." [401]. From the *Christian Century* [92O]: "The book is exciting reading. The two Wellses (H. G. and Orson) might have collaborated to conceive it"; actually the whole review is written in a tone of ridicule and includes such statements as "Why does a collection of such arrant nonsense get such a wide reading . . .?" [115].

Damon Knight wrote about the affair in an apparently detached tone, though somewhat more pro Velikovsky than con. He related the story of Menzel's calculation that "the charge on the sun . . . would be [if Velikovsky were right] 10^{19} volts," and of Bailey's hypothesis on the basis of which he "calculated this charge at 10^{19} volts" [191:140]. How many readers could detect in this that Knight had no business discussing this particular point, since he was using terms as foreign to him as Swahili is to most of us? Twice on one page he talks about *charges* calculated in *volts.* Volts are units of potential or voltage, not of charge, nor can one convert the one into the other without knowing other parameters of the particular system (the electrical capacity). Knight's statement is like saying that the height of a man is 150 pounds or that the money in the U.S. treasury is 10^{19} pints. Similarly, in what otherwise appears to be an informed, judicious, and critical analysis of new religious cults, I was taken aback to read that "it is possible to detect the presence of minute electrical currents, measured in microvolts, in all living nerve cells . . ." [104:198]. One can talk of currents measured in micro*amperes* or of potentials measured in micro*volts,* but currents in microvolts has no meaning.

Altogether, then, if one accepts any piece of writing at face value, one does so at one's peril. Even when many different writers make the same point, that is no assurance that the facts are right—as with Velikovsky's supposed advance claim of the retrograde motion of Venus. Nor is it an assurance that the significance is straightforward if the facts themselves happen to be right—as with Velikovsky's ten years spent on research, actually a few months or years of jumping from one conclusion to another, followed by many years of selecting appropriate references and quotations to bolster the conclusions already arrived at. There is simply no substitute for examining the primary sources for oneself.

I have talked here of various devices by which language is used in an insidious, misleading fashion. Related to this is the development of a special jargon by certain groups, and the use of familiar words but with special—implied but not explicated—meanings. Thus the Velikovskians use, in an idiosyncratic manner, such terms as "catastrophism," "uniformitarianism," "heretic," "orthodoxy." Those words have value for them that is emotional and not merely descriptive—catastrophism and heretic are good, denoting being right and courageous and open-minded. The contexts within which these words are used might make the unwary reader swallow the implied value, which at the very least is a matter of opinion. That the words are familiar, then, does not ensure good and open communication between writer and reader; the words may be common to both, but the implications may exert a subliminal effect on casual readers.

Such situations face us all at every turn. It was not clear to me when I first heard of "Right to Life" that this was anything more than a high moral sentiment, shared by all, and certainly no grounds for fierce public debate—until I discovered that this was a code, a symbol, a banner for a partly religious, partly political movement. Again, when I first heard that educators were moving to make special arrangements for "exceptional" children, I rejoiced that at last we were getting back to the idea of quality education for the brightest youngsters—until I found that "exceptional" in this context had been redefined to mean less, not more. I enjoyed particularly the airline stewardess who had explained that our life-jackets would inflate automatically in the water; she continued, "If you desire extra buoyancy, blow in this tube."

Some cogent discussions of this theme in contemporary life are available [26, 270, 282].

So in the Velikovsky business, as in all public grinding of axes, the observer needs to be aware that what is said is not always what is meant. The opposing camps may seem to be talking about the same things, but frequently (if not usually, even always) tacit assumptions on both sides make the argument look like "two ships passing in the night" [117]. That is one reason why such arguments can drag on interminably. There is no common ground of tacit assumptions or beliefs to which the arguments can be referred for eventual resolution. In the Velikovsky affair there is neither explicit nor tacit agreement as to what constitutes evidence, what can be proved by certain types of evidence, even what scientific activity consists of; yet both sides use the words "evidence," "proof," "scientific."

Communication becomes even more difficult as special new terms are coined: "In-groups often develop special languages, new systematic terms . . . a special language promotes in-group solidarity and out-group ostracism. The priests can talk of special things because they know the words . . ." [34]. Thus to the Velikovskians "collective amnesia" is an established phenomenon; for nonbelievers it is at best a speculation. Consequently, all discussion on that point has been futile, since it was not directed at attempts to demonstrate the effect to the satisfaction of the nonbelievers. Again, de Grazia has introduced a number of neologisms whose use implies the acceptance of Velikovskiana: "revolutionary primevology" [74], "paleo-aetiology," "homo sapiens schizotypicalis" [351], and "paleo-calcinology" [76]. Use of these words marks one as a member of the in-group but makes productive discussion with outsiders more difficult.

The Numbers Racket

Clearly one can be more precise, scientific, truthful, by using numbers rather than words. If a banker tells me that I can get a reasonable rate of interest on my deposits with him,

I do not feel as secure as when he defines that as, say, 5 percent compounded daily. In the absence of the number I wonder what he means by "reasonable." Perhaps he regards 2 percent as reasonable, since I am doing nothing to earn the interest; I might regard 8 percent as reasonable, since I suspect that he lends the money out again at substantially more than that. Though I might be told that I can purchase something for "only pennies a day," it strikes me differently when the bills come in and I find that 90 pennies each day is well over $300 a year.[3] To my friend, "rich" means not worrying about where food, clothing, and shelter are coming from; to me, it means a private income large enough to obviate working for a living.

So there is no doubt that when exact numbers are used, we all know precisely where we stand. We can communicate in numbers without the dangers inherent in the use of words alone.

"What are the mathematical odds against Velikovsky's making so many correct deductions from 'false premises'?" asks Juergens [174]. I note only in passing the common ploy of the rhetorical question, left explicitly unanswered but followed by words intended to convey that there is only one possible answer. Leave aside also how many is meant by "so many," how "correct" they in fact may be, and the lack of general agreement on all those matters. The point to be made here is that Juergens is appealing to the force of num-

3. This trap can so readily be laid because of a monetary system with only two units (cents, dollars), which differ by so large a factor (100). When I moved from Australia to the United States, it occurred to me that the easy-spending American ways (as compared to those of Britons and Australians) might be due not only to differences in the actual standards of living but also to the differences in the monetary systems. In my formative years we had pennies, shillings (12 pennies to a shilling), and pounds (20 shillings to the pound). We may have been rather free in spending pennies—but only up to about 10 of them, because we knew that 12 (a shilling) was "worth something," whereas pennies were not worth much. No one could have told us that 90 pennies was only pennies, because that was 7 shillings and a half, and was therefore by definition appreciable rather than negligible. Proof for this thesis has since come to hand—Australians have become less frugal since the American system of only two units was adopted in Australia.

bers, to the precision that derives from using mathematics; the unwary reader might not know that these "mathematical odds" cannot be calculated because there is no basis on which to make such a calculation.

Velikovsky, however, did not hesitate to answer a similar question with some precision. His reconstruction of history involves so many correlations that the odds against it being spurious are "a trillion or quadrillion against one" [410:339–40]. How reassuring! But ought we really to accept that number (or that range of numbers) without being shown details of how the actual calculation was carried out (assuming that a calculation was actually made and that the numbers "trillion" and "quadrillion" were not simply pulled out of a hat)?

Poor Margolis [236] wrote an article containing 54 errors [72]. This number is quite precise: it means between 53.5 and 54.4. If we knew only that there were between 45 and 54, we would say "about 50"; or, between 51 and 59, "about 55"; saying "54" means not one more and not one less. We have been reminded often that Margolis was wrong in 54 ways [72, 168, 173, 230, 468]. How devastating! I looked at Margolis's article [236] and at the critique of it by de Grazia [72], who had counted the mistakes. Then I played with some numbers myself: in 4 of those 54 instances, Margolis was right and de Grazia wrong; another 4 "errors" were *ex cathedra* statements, 3 of which could be classed as correct if Margolis had given the appropriate references to the literature; 1 "error" was a misspelling; 8 involved nothing more than the use of sarcasm and value-loaded adjectives; 8 were claimed misrepresentation on minor details and 4 on major issues; the remaining 25 "errors" were simply differences of opinion, disagreement with the thrust of Margolis's argument or with his interpretations. So there were indeed "54" —but not errors. Clearly, it is not the number that lies but the word attached to it.

De Grazia also

> attempted a crude analysis of the contents [of a collectively written review [214] of *Worlds in Collision* and] . . . emerged from this little exercise with 27 statements purportedly de-

> scriptive of the work, 4 purportedly empirical statements, 12 purportedly logico-empirical statements, 27 dogmatic-authoritative statements and 8 statements dealing with the character of the author and publisher. A separate summing-up of the evaluative loading of each statement resulted in a total of 2 favorable sentences, 31 neutral sentences and 46 negative statements about the work. In the Velikovsky case, then, rationalistic criticism was heavily subordinated to dogmatic-authoritative criticism of a negative character. This kind of material, if pursued through the Velikovsky case and also through many other scientific case studies, might lead to a complete overhaul of the machinery of scientific evaluation. . . . [73]

It might indeed; reviewers would no longer need to concern themselves with the substantive merits of the work being evaluated but would merely need to make sure that their statements were empirical and not negative. No doubt many authors would be glad to have their books and articles reviewed by that method.

Juergens [72] supported de Grazia's demolition of Margolis by counting elisions of quotation marks, capitalization of words, and similar effronteries. Jueneman [168] was able to top that easily; he reported that "Prof. Lynn Rose of the Dept. of Philosophy, SUNY-Buffalo, read Asimov's 'CP' and counted a near-record 134 errors of fact and logic. . . ." Perhaps it happened to the right victim: Asimov himself [16] referred to "hundreds of places where Velikovsky is wrong," in "at least fifty more passages . . . from [*Worlds in Collision*]," but he failed to specify by page or quotation any of those "hundreds" or "at least fifty." Perhaps Asimov didn't bother because his estimate was so conservative; after all, Menzel [247] knew of "thousands of other erroneous suppositions and conclusions." What an effort must have gone into counting all of them.

One reviewer of *Worlds in Collision* was confident that 90 percent of the facts in it would stand up [92L], but he did not say which were the 10 percent that would remain (or become) supine. Another reviewer was less precise: he could state only that publication of the book had set the American publishing industry back by "at least" 25 years [245]—conceivably, no doubt, by as much as 250 or 2,500 years.

Real precision came, as one might have hoped, from the editor of *Science*. He assured us [73] that "at least half of Velikovsky's ideas have been proved wrong"; unfortunately, he did not list them for those of us who had not gone to the trouble of counting them for ourselves, and who had failed to come across the articles containing the mentioned proofs. Perhaps to make amends for that omission, he assured us that "science can exist and is useful because much of the knowledge in it is more than 99.9 percent certain and reproducible. . . ." I wish that I had been taught how to make calculations of probability to three significant figures in such a context; then again, perhaps that would not be so useful as the ability to judge what portion of the knowledge is contained in the "much" that is so certain. There I could perhaps be helped by the individual who knew that, if not he himself, then "a competent astronomer" at least "could rip 75 per cent of Velikovsky's material to shreds in half a dozen pages of calculations . . ." [226].

It may be that I found these numbers so enlightening because I am not accustomed to seeing this kind of precision applied to such everyday matters as the number of wrong statements made by somebody. I feel much more at home when mathematics is employed for mensuration, for instance. So I was delighted with Stecchini: "the value of the ancient Roman foot had not been estimated much more precisely than 296 mm. . . . But, using the archaeological reports available today, I have determined that the exact value of the Roman foot can be computed as 295.954 mm., and hundreds of tests with different types of data have confirmed this figure . . ." [379]. Of course, I would not dream of questioning that this value "can" be computed (though I do regret not having been shown the computation). But I am very curious about the tools used by the Romans, which enabled them to measure lengths reproducibly to 1 part in 6,000. I had thought that such precision was possible even nowadays only with comparatively sophisticated apparatus. But, then, I had also not known (until I read Stecchini) that "it can be demonstrated that the units did not change one-eighth of an English grain or one-hundredth of a millimeter through millennia. . . ." I wonder whether that is known even at the National Bureau of Standards?

Ad Hominem

Harrison Brown, a consistent critic of Velikovsky's works, wrote: "It is difficult to condemn a man who is searching for the truth, and in spite of his unfortunate approach to scholarly inquiry, Velikovsky is clearly such a man. . . . Even though his conclusions may be as wrong as wrong can be, it is certainly not wrong for him to publish his theories and his viewpoints—provided, of course, that he is willing to subject himself to ruthless criticism" [38]. Why even speak of condemning a man in this context? I could concur with Harrison Brown if he would rephrase the passage to make clear a very important distinction: subjection to ruthless criticism of a man's *ideas and viewpoints* is quite a different matter from ruthless criticism of the *man;* condemnation of those views need not imply condemnation of the individual who holds them.

Even the possibility, let alone the desirability—I would even say necessity—of making that distinction is not widely acknowledged; certainly it is not often acted upon. And that failure is a partial reason for the persistence of violent emotional conflicts; criticisms become arguments *ad hominem,* are intended or taken personally. Throughout the Velikovsky affair harsh denunciations and accusations of unethical conduct were addressed at many individuals and groups: at Velikovsky himself, at those who gave summaries of *Worlds in Collision* before the book was published, at Velikovsky's critics, at publishers, reviewers, editors.

We readily assume that those who act unethically should know better, and frequently we say so explicitly. In doing that, we likely fall into error in one or both of two ways. First, we assume that those criticized share our own ethical convictions (or, much the same thing, that there is an absolute ethical code—ours—to which all do or should subscribe). Second, even if the first assumption is correct in the sense that those whom we criticize do indeed have the same ethical standards as we do, we err in talking of the breaches of ethics as though they were deliberate and intended, which in fact is rarely if ever the case. Our ethical views of particular situations result from the making of judgments, a fallible pursuit. We find it difficult to conceive that another's judgment could *honestly* differ from our own, when in fact that is the reality more often than not. For example, I have emphasized Velikovsky's misrepresentation of certain sources, misrepresentations so plain as to make effectively for untruth. I suggested that Velikovsky either misunderstood completely what he read or misrepresented deliberately. I am firmly of the opinion, actually, that he misunderstood and had no conscious desire to misrepresent. Velikovsky, after all, sincerely believed that he was right and, as a corollary, that many others were wrong, including scientists about important technical details in their fields. Given that belief, naturally Velikovsky chose to quote only those things that appeared to him to be "correct," namely, what fitted his own ideas. So when he quoted out of context and thereby altered the meaning, in his view no doubt he was merely quoting what was valuable and sparing his readers the errors in the cited sources. I suggest that Velikovsky was more likely "guilty" of unconscious sophistry than of conscious misrepresentation.

Another example: it is evident that Payne-Gaposchkin misrepresented Velikovsky's contentions in her attempt to show them to be dynamically ludicrous. My guess is that she would have been genuinely horrified at the thought of deliberately misrepresenting anyone's argument; she also likely indulged only in subconscious slippery and sophistical thought. Knight [191:136–37] suggests that since, in her view, Velikovsky was certainly wrong, the details of the argument were not all that important—what was important was to demonstrate to the lay public the *type* of argument that could prove Velikovsky wrong. Nor do I doubt that Menzel's thinking went in analogous fashion. I do not imagine that he would regard it as proper to ask a fellow scientist not to publish something simply because it provided support for an idea that stood in opposition to Menzel's own (as, however, he apparently did [173]). Rather, Menzel was so convinced that he was right and Velikovsky wrong that any article that indicated otherwise must therefore also be wrong in some way.

I have rarely encountered relatively sane people who are moved by the desire and intention to injure others and who actually do so by deliberate and conscious actions that they themselves see as unethical.[4] But I have encountered relatively many who caused harm to varying degrees as a result of naivety, incompetence, personal feelings of insecurity, inability to learn from experience, unconscious sophistry, and the like—given the chance, we are all guilty of that sort of thing. Strong emotion makes it more likely that one will behave—contrary to what one would like—in such a manner, which is a good enough reason for attempting to come to terms with reality and to avoid the outbursts of anger and frustration that result when we say to ourselves that something "should" be so when it quite obviously is not.[5]

In my view, those who suffered most tangibly in the Velikovsky affair through no fault of their own were Putnam[6] and Atwater (who lost their jobs) and the Macmillan

4. There is, however, one individual whom I knew quite well and whom I put into that category, no doubt because I was too close to the situation to be objective, having myself languished under his authority for some time. Speaking of him to a close friend, I drew a comparison with those who had served as guards in Nazi concentration camps. My friend, who had spent several years as a prisoner in the Belsen camp, set me straight: "anyone," he said, "is quite capable of being a guard in a concentration camp." Those were people no different from other people, and the evil was not deliberately designed as such by individuals who saw themselves as acting unethically or immorally.

5. Techniques for recognizing and avoiding the irrational "should" and the ensuing irrational anger have been proposed by Ellis [101, 102] and Maultsby [239].

6. But Putnam was given a year's severance pay, and almost immediately found a new position, with World Publishing Company. Moreover, it seems that the president of Macmillan had not been happy with Putnam in any case and seized a convenient opportunity to dismiss him [99A].

Company. Much of the outrage expressed by Velikovsky's critics was directed at the publisher [37, 38, 60, 112, 185, 214, 269, 393]. Macmillan was guilty, it was said, of not having scientists evaluate a manuscript that purported to be in some ways scientific, of advertising as "science" a book that was not, and of using wide publicity in an effort to achieve large sales and make money for the author and the publishers. At the same time several critics pointed to Macmillan's previously excellent reputation for its list of scientific texts, supposedly making publication of *Worlds in Collision* yet more heinous.

I would have concluded that a publishing house with an excellent reputation which published a hoax or a work of charlatanry or pseudo-science had been misled and made a mistake; it would not have seemed at once obvious to me that Macmillan had changed its publishing policy in order to make money as rapidly as possible irrespective of the means employed toward that end. Yet the complaints of the critics indicate that many of them jumped to that conclusion.

In fact, Macmillan had followed the normal procedure of having the manuscript of *Worlds in Collision* evaluated by outside readers. We know the identities of two of them: O'Neill, science editor of the *New York Herald Tribune,* and Atwater of the Hayden Planetarium and the American Museum of Natural History. Surely, the positions these men held would make them seem appropriately qualified to pass judgment on a book that dealt in part with astronomy as well as other sciences and which was written for the general reader. Moreover, the reviewers recommended such changes in the manuscript as deletion of an earlier catastrophe [166]. To Macmillan's editor, Putnam, this would have been a clear indication that the reviewers were doing a critical and conscientious job. In retrospect one can perhaps make the judgment that Macmillan's choice of reviewers was unfortunate, but there is no evidence at all that they did not make an honest attempt to obtain informed and authoritative opinions.

Further, when Shapley protested before *Worlds in Collision* was actually published, Macmillan sought reviews from three new and impartial referees. Apparently, although the book was already in press, the publishers were prepared to take a financial loss at that stage rather than to publish something that was obviously pseudo-science. Again, we do not know the identities of all the new reviewers, but one of them reportedly [172] was the chairman of the Department of Physics at New York University. That would seem to be a thoroughly sensible choice. And, having chosen individuals whose judgment could be presumed to be both informed and impartial, Macmillan accepted their judgment: a two-to-one majority in favor of publication.[7]

I fail to see how the publishers could have acted more responsibly. In the end, having accepted the best advice available to them, they naturally proceeded on the assumption that what had been found acceptable as popular science by qualified referees could legitimately be advertised as such. I trust no one will argue that publishers ought not to seek the widest possible sales for their books once they have reached an honest conviction that the books themselves are honest ones. A publishing house that does not make a profit will shortly cease to be of use to anyone. There are respected houses—Macmillan, for instance—that strive to publish works of intellectual and artistic merit, and in the process lose money in a number of cases. The ability to continue the attempt at quality in publication depends on making a profit somewhere in the line, and a compromise has to be made with what the widest public wishes to read. Fewer and fewer publishers can afford to take chances on books whose sales

7. The reviewers' letters, now available in the Macmillan files at the New York Public Library, reveal that none of the three actually recommended against publication [99B]. In addition to C. W. van der Merwe, professor of physics at New York University, the readers were Clarence S. Sherman, associate professor of chemistry at Cooper Union College, and E. M. Thorndike, head of the physics department at Queens College. Though there are no statements against publication of the book, two of the three letters contain strong criticisms on some points of science. (The Macmillan files also reveal that similarly strong criticisms on points of science had been made by O'Neill, who was one of the first readers of the manuscript originally submitted by Velikovsky to Macmillan.)

are likely to be limited, no matter what great intrinsic merits those manuscripts are perceived to have. As a result, scholarly authors are forced increasingly to turn to university presses, which require subsidies from the institutions, or to "subsidy" or "vanity" houses—publishers to whom the authors themselves pay the cost of publication, very rarely recouped by the authors from subsequent sales. (A most enjoyable account of subsidy publishing has been given by one of the leaders in that field, Edward Uhlan [405].) As I was writing the first draft of this book, I found in my daily newspaper a column [150] that seems germane here. It was entitled "Publishers Undermining Status of Being an Author":

> There was a time still warm in memory when the name on a book meant something. Book publishing was a respected business and authors wore the halo of achievement.
>
> But talent no longer is a requisite of publication, and more and more bad books are being written. . . .
>
> Instead of ability, what seems to stir the interest of publishers is the name or an association or the incidence of notoriety and impropriety which the so-called author can bring to a book. . . .
>
> Of course bad books and bad writers are nothing new. What is new are the homage and deference they are receiving in the publishing trade. . . .
>
> Low-grade "literature" used to be the means to an end. An author dashed off a potboiler and a publisher published it, both with the same thing in mind—to carry them through for something worthwhile.
>
> There has been a complete reversal—what once were stopgaps are today's "achievements." Their authors, or non-authors, are wined and dined and pampered, while legitimate writers of quality are pretty much ignored as stepchildren.

I sympathize with the sentiments expressed, but how well do they stand up to analysis? Take a "hack" who makes money for himself and for his publishers because so many of his fellow citizens enjoy reading his books—should his publishers hold him at arm's length and not wine and dine him?

Should deference be reserved for those whose work appeals only to a small elite? Was it ever really true that "bad writers" whose books sold well were *not* given "homage and deference . . . in the publishing trade"? One cannot hold publishers responsible that "notoriety and impropriety" make for large sales. Our columnist simply wishes that our pluralistic society were other than it is, and, translating that wish into an irrational "should," he inevitably waxes indignant.

15

Some Realities about Science

> He had been eight years upon a project for extracting sunbeams out of cucumbers, which were to be put into phials hermetically sealed, and let out to warm the air in inclement summers.
>
> —Jonathan Swift

The Velikovsky affair demonstrates that misconceptions about science are held not only by nonscientists but also by scientists. Although some aspects of scientific activity inevitably become familiar to those who practice it, scientists commonly are very unself-conscious about the assumptions underlying their actions and about the philosophic implications of their modes of procedure. The question "What is scientific truth?" does not usually bother the scientist; that question is left to the philosophers and is seen by most scientists as irrelevant to the actual pursuit of science. "Most writers on science now [1967] accept the validity of science as unquestionable and neither in need of philosophic justification nor capable of justification . . ." [321].

What Is Science?

Much indeed has been written about this. Not only do I find lack of agreement as to conclusions, but I believe that the very attempt to give a concise definition must inevitably lead to error. I shall proceed on the basis that science comprises all those matters that are commonly called "science," and that we are interested in realities about that. In other words, I shall attempt a purely descriptive approach as the only way to avoid misconceptions. When one tries, as many have done, to construct a model or theory of what science is, one isolates and focuses on certain characteristics that are regarded as "the" important ones. But scientific activity is so diverse that inevitably exceptions can be found to any generalization. The danger of the theoretical construct is that one may get so enamored of it as to come to believe that the model describes how science should be carried on or what science should be. Such an attitude inevitably clashes at some point with what actually takes place, and thereby leads to frustration, anger, and unresolvable arguments.

So I begin by noting that no satisfactory, global, concise definition of science is possible. One cannot define it in terms of experimentation, since there are *observational* sciences as well as *experimental* ones. Science is also not quantitative as opposed to qualitative, for much good science is not quantitative at all (a distinction is sometimes made between so-called *hard* and *soft* sciences, the former ones being the more mathematically and quantitatively rigorous). We speak of the *natural* sciences in contrast to theology or moral philosophy; where does epistemology fit? Is it part of science? There are the *physical* and the *biological* sciences. We talk of *modern* science and of many other kinds of science.

One can distinguish three different aspects of science (and people will often talk as though science is merely one of those aspects). In one sense science is a particular body of knowledge, more precisely a set of phenomena accepted as facts and a set of theories accepted (at any given time) as "explaining" those phenomena: a collection of what Kuhn [204] has called paradigms. In another sense science is defined by a particular approach, the scientific method (though there is no agreed definition of exactly what that method is); the actual knowledge accumulated then takes a somewhat secondary role, its nature resulting inevitably from the method used to obtain it. In yet another sense science is defined by the people and institutions that carry on scientific activity; science is whatever happens to be done by that community.

Not only does there exist the misconception that science can be equated with a single one of these aspects, but there are also many misconceptions about each of these aspects themselves—what scientific knowledge is, what defines scientific activity, and what sort of an institution science is.

Scientific Knowledge

Common to many inside and outside the scientific community is the misconception that scientific knowledge is the same as truth. In common usage the adjective "scientific" has come to assume the force of certainty, reliability, truth.

Payne-Gaposchkin [287] was displeased that "this scientific age" is uncritical, revealing a belief that a scientific age (whatever that might be) is one in which error has become an aberration because of some high degree of critical reasoning; that some attribute of science makes "scientific" synonymous with "not uncritical," not easily misled; that "scientific" means "not wrong," uncomfortably close to saying it means "true." Again, Velikovsky's work was said to be "divorced from scientific reality" [257], as though "scientific" reality were some special, warrantedly reliable reality different from ordinary reality. Reviewers wrote of Velikovsky's "scientific correlations" [217, 218] and "scientific evidence" [378]—palpably unwarranted uses of that adjective in a pragmatic sense, because Velikovsky did not work as a scientist. Worse, the adjectives do not merely imply that specific untruth but give the even more broadly false impression that the correlations and evidence are in some manner definitely true and valid.

I can choose to say, "it is a fact that the sun will rise tomorrow" or "it is a scientific fact that the sun will rise tomorrow." In most company I would be offered no argument against either formulation. Yet, as I pointed out in Chapter 6, even the use of "fact" here is, strictly speaking, unwarranted;

insertion of the adjective "scientific" implies that scientific expectations are somehow more reliable than nonscientific ones. In common usage nowadays "scientific" is synonymous with "true"; we habitually forget that science does not deal in absolute truth; we use semantic devices that obscure the distinction between science and truth; we talk as though they were one and the same thing.

Thus Larrabee [211] saw as the issue in the Velikovsky affair whether a natural fact could be uncovered independently of science. That is not an issue at all: it is a question that can confidently be answered in the affirmative as soon as it is posed. Larrabee suffered from the misconceptions that scientific activity is synonymous with the uncovering of natural facts and that natural facts will inevitably (let alone quickly) be incorporated into science; again, that science, fact, and truth are the same thing.

One can distinguish two questions about the relation between scientific knowledge and truth. First, is it inherently possible to discover truth via science (or, more generally, by any other means)? Second, if so, to what extent have we already succeeded? The first question is in the realm of epistemology, and I leave it to the reader to delve as far or as little into that as he wishes. I give only my personal opinion: in practice we have no direct means of experiencing whatever absolute, external truth and reality might be—we interpret various sensations and stimuli and form mental constructs of where those sensations and stimuli might have come from. In science we follow the same procedure, albeit using quite complicated devices and techniques. In my view we can never *know,* for example, what an electron *is;* we merely find mathematical expressions whose manipulation gives certain correlations, and when experiments produce similar correlations, we have a scientific truth—electrons have both wavelike and particlelike properties. A scientific truth is a limited kind of truth; applied to situations similar to those in which it was discovered, a scientific truth is very reliable.

However, even if my view on this is not accepted, the second question surely leads to the same conclusion. An absolute truth is final, unchangeable, irrevocable; no contemporary scientific truth is known to have that status, is even claimed to have that status. Rather, all experience indicates that scientific laws and theories need some modification as time goes by. The laws of conservation of energy and of mass were, for a goodly length of time, regarded as true because they were very reliable. Doubtless there were scientists who believed that here, at least, were a couple of laws that would never need to be modified. But Einstein suggested that these truths were limited ones, in that energy and mass were interconvertible—as was later demonstrated in nuclear fission and fusion. So those two laws became a single law of conservation of energy-mass, whatever that might be, and there are doubtless some who believe that here is one law that will never need to be modified. Even beyond that, the existence of conservation laws of many types—angular momentum, energy-mass—led in the past to the unspoken assumption that all similarly fundamental entities would also be conserved. This belief is now seen to be false following the demonstration (in 1956) that "parity" is not conserved.[1]

It is surely unlikely that we have reached a final truth, never to be modified, in any area of science. For that reason alone one had better view scientific truths as limited ones, applicable in limited ways. In the Velikovsky affair many words were spent that do not make sense if one accepts this view. The controversy was waged on both sides about

1. This matter illuminates several facets of science [28:52]. First, the belief that parity is conserved was held solely on grounds of analogy, because other such quantities appeared to obey conservation laws; there was no direct experimental evidence at all that parity itself was conserved. Second, experiments in whose results the *non*conservation of parity was implicit were performed in 1928–30, but the conclusion of nonconservation was not drawn from those results because it was, at that time, a premature discovery (see below)—nonconservation was not yet scientifically conceivable. Third, scientists lack a self-conscious, critical attitude in the normal course of events, and may not even be aware that they believe something for which there is no evidence. Yang, one of the individuals who was awarded a Nobel Prize for the discovery in 1956 of parity nonconservation, said, "That parity conservation in the weak interactions was believed for so long without experimental support was very startling."

whether Velikovsky was right; implicitly accepted was the notion that science could answer that with certainty. All that could have been established was that Velikovsky's ideas were at variance with reliable existing scientific truths, which Velikovsky had admitted at the outset anyway. Further, it could have been shown that Velikovsky's concepts were not useful as a basis for disciplined scientific activity. But one cannot by science absolutely prove Velikovsky wrong, just as one cannot by science disprove the proposition that the earth and all else were created in 4004 B.C., complete with fossils and other apparent indications of evolution over billions of years. Both notions may seem to many of us to be highly, even absurdly, improbable—but, then, so is life itself, and the universe, after all. One person's faith is another's improbability. As Boring [34] observed, "Functional, practical truth is social, the truth of agreement, and . . . absolute truth is actually not available in science. . . ."

Moreover, scientific truths, though useful in concrete ways, are not a viable basis for human activity as a whole; as one scientist has put it [28:108], science is irrelevant to "the underlying metaphysical and moral truths by which one lives. . . ." The nineteenth-century belief that science could invalidate religion was as wrong as the earlier view that science could validate religion. Science and religion are simply different parts of the human experience. Each individual is free to choose what sort of relation he makes between them, but no necessary relation has been objectively demonstrated, nor can one be. The view [18] that Velikovsky's work might serve to link science and religion is a disservice to both, based on a misconception of what these things are.

Recognizing that scientific truth is of a limited kind, let us not then forget, however, how remarkably reliable and useful it can be within its own sphere. The comment [274] that our high standard of living bespeaks the correctness of our basic ideas about the physical universe is admittedly naive, but application of existing theories has made possible some remarkable things. Landing men on the moon (more important, bringing them safely back) demonstrated that what science knows may not be absolute truth but it is a very powerful, usable, practical truth. Over the long haul, scientific truths can claim reliability. Laws that do not work in practice are eventually discarded, and those that remain in use are pragmatically justified. It was quite naive of Velikovsky and his supporters to claim that Velikovsky was pragmatic, since he stuck to the "facts" even when they contradicted the "laws" of science: those laws themselves are pragmatic, built from many past observations and experiments.

Scientific Facts

Granted that the whole absolute truth is not to be found in science—are there not at least some scientific facts that are absolutely true? Well, perhaps so, particularly if past experience is an infallible guide to the future, and if we have certified infallible evidence about what happened in the past. But it turns out, as we try to think of examples, that the more reliable a fact is, the less interesting it is, the more trivial, the closer to being a tautology.

It is a fact, for instance, that the atomic weight of naturally occurring phosphorus is 30.975, but that merely says that the weighted average of those atoms has 30.975/12 times the mass of the atom of carbon-12, *to five significant figures*—the *exact* (absolute) value is unknowable. A set of experiments designed to measure that value will always give varying results, because of the finite precision of any technique and through the occurrence of "chance" "experimental" errors. Even if the exact value were somehow obtainable, this in itself would be of rather trivial interest: it becomes significant only when we infer that the most common type of phosphorus atom has 31 "particles" in the nucleus whereas the carbon-12 atom has 12. The interest, the significance, increases as the fact is correlated with other facts, so that generalizations emerge about the structures of atoms. Even when we talk about the importance of scientific facts, what we actually regard as important are the correlations, laws, theories that bind facts together.

Facts alone have little *scientific* value. (They may, of course, have great value from other viewpoints—that aspirin is an analgesic is of great medical value but little scientific value, since we do not understand how aspirin works and consequently cannot link this fact with other facts.) The accumulation of concordant facts is scientifically welcomed. One can then extrapolate—predict other, similar facts—and if the extrapolation receives support from experiment or observation, a law results. Laws are shorthand statements that express the salient common features of a set of facts. When several laws can be connected in some way, the power of generalization is again increased, and one then speaks of theories. Once theories are established, one commonly uses them in lieu of the underlying facts—one takes the theory for granted, regarding the laws as deducible from the theory and the facts as deducible from the laws.

Thus scientists come to ascribe more truth than is warranted to laws and theories. At the same time, as a result of experience they are extremely cautious about what is accepted to be a fact, and what significance can be attached to a single fact or to a small number of facts. Almost always the "same" experiment gives different results each time it is carried out; the variations are often quite small, but in some cases they can be large indeed. If one carries out the same experiment often enough, sooner or later there will come a result that is very different, because it is simply not possible to control all the variables perfectly. Every experimenter has had experience of discordant facts—occurrences that are completely outside the expected range, for no apparent reason. Most experimenters have had the experience of a discovery that could not afterward be reproduced. Few scientists have not published papers in which errors of fact were later demonstrated. That sort of experience makes scientists (as always, of course, with some exceptions) reluctant to propose modifications of established theories merely on the basis of some discordant and apparently unexplainable facts. Since laws are the expression of many facts, and since the uncertainties associated with individual "facts" are well known, it is common practice to discard or ignore the discordant facts until such time as the "discordant" ones come to predominate over the concordant: then, only then, are the laws or theories questioned anew.

That, in part, is why Velikovsky's evidence did not impress the scientific community. His historical "facts" depended, for their significance, on how they could be correlated with other facts, how they could be interpreted in terms of a theory. Yet the "facts" themselves were but interpretations, based on such assumptions as that there are reliable accounts of real events at the bottom of myths, folklore, religious texts. Scientists, as a result of sometimes bitter experience, are wary even of facts obtained under strictly controlled conditions with inanimate objects of well-defined character; inevitably they give little credence to eyewitness reports, still less to purported eyewitness reports handed down through many generations, sometimes merely orally. Velikovsky's judgment of what facts are, and how reliable they can be, differs in this from the norm of scientific judgment. Had Velikovsky's interpretations been in some way consonant with accepted laws or theories, they would not have been rejected outright; but, being contrary to a whole set of ideas that had been satisfactory in the past and continue to be satisfactory (except in relation to Velikovsky's scenario), Velikovsky's "facts" could not be taken seriously by scientists.

Scientists have great confidence, for which there is some justification from past experience, that—eventually—discordant facts will become concordant as knowledge grows, new factors are identified, theories become more general. So they are used to living with a host of unexplained matters—extinction of the mammoths (and dinosaurs, and others), periodic reversals of the earth's magnetic field, the origin and differentiation of languages, the mechanisms responsible for the Ice Ages—all the unexplained things emphasized by Velikovsky. Those are just "lonely" facts; they have not been correlated or fitted in, so they have little scientific value. Velikovsky sought to turn the scale of values upside down, from a scientist's viewpoint; he placed greater importance on the outcast facts than on the set of concordant ones.

Velikovsky's advance claims are also but lonely facts. Venus is hot—so what? Since we have no satisfactory theory to account for the existence of the solar system as a whole, we are not much bothered by *any* data concerning physical conditions on the planets, and we have a definite preference for data that can be correlated, that will give clues to a theory that can fit all the planets. Velikovsky's chain of reasoning is not an objective one as regards Venus's temperature (and the other advance claims). He does not give even approximate numbers to the amounts of heat supposedly involved when Venus left Jupiter, encountered earth and Mars, and came close to the sun, so there is no way to assess objectively the plausibility that some 27 centuries later the temperature would still be "high" as a result of those earlier events. Indeed, another individual could equally well use Velikovsky's scenario to reach the conclusion that Venus should now be "cold": it must have radiated heat very strongly when it was hot, and also could not have settled into its present orbit without dissipating a great deal of energy; some unknown, very powerful mechanism was responsible for this dissipation of energy; therefore most probably much heat energy was dissipated as well as kinetic energy, so the planet should now be cold.

To make science, it is not enough to connect new "facts" with new "theories." Somewhere a connection has to be made to what already exists in science, and Velikovsky failed to make that connection. His facts and theories are both *ad hoc,* limited in their application to the phenomena explicitly dealt with. That is not particularly useful or valued in science. On the other hand, some of the dangers in the scientific approach are also evident from the Velikovsky affair. That a number of facts can be correlated is easily taken to mean that there is involved some rather direct, albeit yet-to-be-discovered cause, and the correlation can come to have virtually the weight of an established theory. So Bode's law, a purely empirical formula (as Velikovsky rightly pointed out), was taken by some scientists as evidence against Velikovsky's scenario; when a theoretical basis for Bode's law was later found, it turned out to make Velikovsky's scenario, if anything, *more* rather than less plausible (Chapter 5, *Scholarly Support*).

Scientific Theories

I pointed earlier to the connection among facts, laws, and theories: concordant facts lead to laws, correlation of laws leads to theories. That sequence, however, is one of logical connection and not of time. Theories may be proposed on the basis of little or no factual evidence, but they come to be given credence only to the extent that they can then be correlated with facts or with other laws or theories. In a sense theories come to "replace" the facts that originally served to validate them. One learns the theory rather than the facts, since that is so much more efficient—it is a mnemonic device. As a science matures, the theories become progressively more powerful [373:158], and the reliance placed on them can easily appear to be an instance of hidebound dogmatism and orthodoxy. Yet that is actually not the case: the theories are shorthand for a vast array of facts, the overwhelming majority of known, connected facts. The Velikovskians allege that scientists adhere to uniformitarianism and the law of gravitation as a matter of dogma. Not so—they do so because of the overwhelming evidence that those ideas are reliable and thereby reflect an important part of reality.

Conant [55:165, 173] distinguishes speculative ideas, broad working hypotheses, and well-tested conceptual schemes. In science the speculative ideas will be ignored or rejected unless they connect in some acceptable manner with existing knowledge or produce ideas for new experiments—that is, unless they have some perceived scientific value. What might seem outrageous speculation is countenanced if there is some scientific value in it. So, at a time when light was "known" to be wavelike, Planck was able to have published a speculative paper showing that the actual known mode of emission of radiation by a hot black body could be derived from equations that treat the radiation as being particulate rather than wavelike. That speculation could be entertained because it fitted with a frequently observed fact that had been bothersome to a considerable extent—not merely unexplained but unexplain*able* on the basis of existing notions. Moreover, Planck speculated that the particulate emission might result from the properties of the emitter, so that he was not really challenging the existing orthodoxy that light "is" a wave (or, at least, he put it in such a way that the conflict with the known was not inevitable). Planck based unorthodox ideas on well-established results obtained by established methods, but not in a way that contradicted all existing theory. How different was Velikovsky's approach—he challenged the laws of Newton and the ideas of Darwin at the outset, and showed no inclination to attempt to fit his ideas with any part of established knowledge.

De Broglie was perhaps even more speculative than Planck. If what we thought of as wavelike also had particulate properties, why not the other way around? Perhaps electrons, for example, have wavelike properties? Certainly a fascinating idea, and properly to be valued as original. At this level, however, it would have been of no *scientific* value, and would have been ignored by the world of science (though perhaps taken up in the world of science fiction). But de Broglie went further, and calculated possible wavelengths for electrons in atoms—and, lo and behold, the results fitted with existing ideas about the "orbits" of the electrons. So the speculation connected with existing facts, and was not only acceptable but even welcome. Further, it suggested the idea that beams of electrons might show the wavelike property of diffraction, a quite feasible experiment which, when performed, indeed showed diffraction.

De Broglie gave reasons based on existing ideas for expecting to find diffraction of electrons. Velikovsky's requests for experiments were quite different: take my ideas, he said, which are not only not connected with, but actually contrary to, existing knowledge, and look for such-and-such an effect. No scientist would waste his time on such an endeavor; there are innumerable other tasks that are more likely to produce something of scientific value.

Successful, creative scientists combine originality, the sparking of new ideas, with the ability to analyze and criticize those ideas before propounding them publicly. Knowing the norms of scientific procedure, a scientist rarely ventures to publish speculations for which he finds no connections to the conventional wisdom; if he does so publish, he is not likely to be surprised at the chorus of hoots and jeers from his colleagues. Velikovsky did so publish, and was then displeased by the foreseeable reaction.

The Velikovsky affair also illustrates that humanists and scientists have very different attitudes toward originality: "new directions and new goals in scientific exploration. Dr. Velikovsky has initiated both and deserves, at the very least, the respect and consideration authentic men of science owe each other" [182]. Stove [453] said that Velikovsky's originality ought to be admired irrespective of any substantive merit that his original ideas might have (see Chapter 4, *Right or Wrong Is Not the Issue*).

Assume for the moment that Velikovsky's concepts indeed were startlingly original (though I have argued otherwise; see Chapter 13, *Velikovsky's Originality*). In science, originality per se makes a reputation for no one and wins no prizes, be it tenure at a university or a Nobel award. In science, what is acclaimed is originality *that also works,* that has connections with what was previously known, that stands up to the test of experiment and of time: "ideas are easy.[2] They are cheap. It is the proving of a suggestion beyond a reasonable doubt that makes it valuable" (see Chapter 4, *Something Amiss in Science*). De Grazia's [72] suggestion that there

2. I. J. Good [120] has pointed out that this statement ought to be qualified: "Ideas are not always easy. Sometimes the idea is nearly all the battle, especially in mathematics."

is a shortage of good wild-guessers remains incomprehensible to me; science has no use for wild guessing.

Larrabee [211] maintained that when a theory results in successful predictions that were previously unforeseeable, that theory is thereby judged valuable. He is wrong in that (pp. 89–90), and therefore also in applying the generalization to Velikovsky's work, because, first, valid predictions from a theory can in principle be made by any competent practitioner of the relevant discipline. Velikovsky's predictions are not of that sort—no one knows how he was led to make them, since he has not explained it (for instance, regarding radio noises from Jupiter; see Chapter 6). No two people—having read *Worlds in Collision, Cosmos without Gravitation,* and Velikovsky's other books—would necessarily make the same predictions (for instance, regarding the temperature of Venus; see preceding section). Velikovsky even contradicts himself on some of his advance claims—whether the petroleum of Jupiter and Venus is biogenic or abiogenic, for instance (pp. 126–27).

Second, it has not been shown that Velikovsky's predictions were "unforeseeable" in the light of conventional ideas—some of them were unforeseen, which is not at all the same thing. Scientists have no doubt that, within the framework of accepted theories, means will be found to explain the temperature of Venus, as well as the temperatures of all the other planets, and the retrograde rotation of Venus, as well as of the planetary satellites that also have such an "anomalous" rotation.

Third, Velikovsky's "theory" does not have the desired characteristic of applicability to similar situations. As indicated in the previous sentence, science seeks not *ad hoc* explanations for conditions on Venus but general explanations from which will follow the temperatures of all the planets and the directions and speeds of rotation of all the bodies in the solar system.

So there are sound scientific reasons for ignoring Velikovsky's ideas. But, of course, *unsound* reasons were also given: for example, that he gave no causative basis for his conjectures [257]. That was rightly countered [456] by pointing out that, although a causative basis may be an advantage, its lack is not a necessary refutation of the theory. Recall in this connection that some scientists were content to accept Bode's law as significant and meaningful in the total absence of a causative basis;[3] there are innumerable other examples in science.

The widespread preoccupation with causation among scientists is quite unfortunate. The language of "because," "for," "since," is a very convenient one, but its use gives the incorrect impression that something is actually being explained, an answer given to "why?" Since science does not deal in absolute truth, it has no answers to the fundamental questions that begin with "why?"; science deals in relations and correlations, and the semantics of causation is used as a shorthand description of correlations. I tell my students that chemical reactions occur, in part, "because" an atomic shell containing eight electrons has great stability. That is certainly not a fundamental reason; it is the expression of pragmatic correlation of many chemical facts, but to phrase it more correctly would be unwieldy. Metals show their characteristic properties "because" the outer electrons are relatively loosely bound to the atoms—again not a fundamental reason: metallic character has been found to *parallel* the ease with which electrons can be removed. We teach established, reliable theories as a more efficient, speedier approach than presenting masses of facts for memorization, and then show how the facts can be "deduced" from the theory. That pedagogic convenience carries a heavy burden of disadvantage, however, since science is thereby taught as something other than it actually is. From this approach, I believe, stems the typically unthinking acceptance by scientists of established views and the tendency to ascribe to those views more certainty than can be philosophically justified.

3. I am indebted to I. J. Good for drawing my attention to the fact that other scientists in another context debated whether Bode's law had significance [121].

Science and Common Sense

In describing the interrelation of facts, laws, and theories that one finds in science, I have perhaps made it seem a quite reasonable interrelation, a common-sense one; in some ways it is. Some writers have gone further, and have characterized science as merely the application of common sense to experience.

That is perhaps not too wrong a description of sciences in early stages of development. Take the physics of gases, for example: the ease with which gases can be compressed makes obvious common sense if one thinks of gases as particles thinly scattered in empty space. That the volume is halved when the pressure is approximately doubled also makes sense: halving the available space—all else being unchanged—will result in the particles of gas colliding twice as frequently as before with the walls of the container, thereby exerting twice the previous pressure. So this kinetic theory of gases fits with common sense, and when a quantitative theory was built on these ideas, it was nice but not unexpected that there was good agreement between experiments and calculations based on the theory.

Now consider the viscosity of a gas. Viscosity describes resistance to flow. Blood is thicker than water—it has a greater density and flows less easily; milk is lighter, less dense, and less viscous than bread or cake dough. The denser a substance is (that is, the more tightly packed together the particles in it are), the less easily it flows—we know that from everyday experience and it "makes sense." Surely, then, the viscosity of a gas will also increase as its density increases, as the particles are forced closer together and interfere more with one another's motion. But, alas—calculations from our otherwise so successful kinetic theory predict that the viscosity of a gas will be *independent* of its density. We do not reject the theory immediately, however, because it has explained so much else; we carry out experiments just to be sure that what "makes sense" actually happens. And then, a surprise: experiments verify the prediction of the theory, contrary to common experience and to common sense; the viscosity of a gas indeed does not change as its density is altered. Jeans [161:164], in a book quoted by Velikovsky, points out that here is an example of a scientific fact that runs counter to common sense. There are many other examples.

Take the bread-and-butter of chemistry, the periodic chart of the elements (Fig. 12), so useful because the properties of the elements show many regular trends along the horizontal *periods* and the vertical *groups*. The most metallic elements are to the left and bottom, the least metallic to the top and right. The atoms of elements lower in any given group lose electrons more readily than those of elements higher in the group. The sizes of the atoms decrease rather regularly as one proceeds to the right in a given period. The electronic structures of atoms, which determine the properties of the respective elements, are mirrored in the periodic chart; for example, in the "A" groups the number of electrons in the outer shell of an atom is the same as the group number.

Since there are so many and such regular trends, it makes common sense to look for regularities in all properties, at least in those that would seem fairly straightforward. But that turns out not to work. Only two elements, mercury and bromine, are liquids at ordinary temperatures and pressures. Bromine fits into a trend, the two elements above it in the group being gases, and iodine below it a solid. But all the elements around mercury are solids.

Again, in Group IA the atoms lose electrons progressively more easily in the sequence lithium, sodium, potassium, rubidium, cesium—the usual trend in all the groups. Now, one can make electric batteries by using as one electrode (the cathode) a material that easily loses electrons, and as another (the anode) one at which electrons are readily taken up. Coupling a given anode with cathodes of various Group IA elements produces a series of batteries, and one finds that the voltages of those batteries increase in the order sodium, potassium, rubidium, cesium, lithium. Lithium, ex-

PERIODS / GROUPS	IA	IIA											IIIA	IVA	VA	VIA	VIIA	
2	3 Li LITHIUM	4 Be BERYL-LIUM											5 B	6 C	7 N ==	8 O OXYGEN ==	9 F FLUOR-INE ==	10 Ne ==
3	11 Na SODIUM	12 Mg MAGNE-SIUM											13 Al	14 Si	15 P	16 S SULFUR	17 Cl CHLOR-INE ==	18 Ar ==
4	19 K POTAS-SIUM	20 Ca CALCIUM	21 Sc	22 Ti	23 V	24 Cr	25 Mn	26 Fe	27 Co	28 Ni NICKEL	29 Cu COPPER	30 Zn ZINC	31 Ga	32 Ge	33 As	34 Se	35 Br BROMINE ≈	36 Kr ==
5	37 Rb RUBID-IUM	38 Sr STRON-TIUM	39 Y	40 Zr	41 Nb	42 Mo	43 Tc	44 Ru	45 Rh	46 Pd PALLA-DIUM	47 Ag SILVER	48 Cd CADMI-UM	49 In INDIUM	50 Sn	51 Sb	52 Te	53 I IODINE	54 Xe ==
6	55 Cs CESIUM	56 Ba BARIUM	57-71	72 Hf	73 Ta	74 W	75 Re	76 Os	77 Ir IRIDIUM	78 Pt PLAT-INUM	79 Au GOLD	80 Hg MERCURY ≈	81 Tl THAL-LIUM	82 Pb	83 Bi	84 Po	85 At	86 Rn ==

← METALS

← METALS | NON-METALS →

ALL THESE ELEMENTS ARE SOLIDS EXCEPT THOSE MARKED ≈ (LIQUID) OR == (GAS)

METALLIC CHARACTER INCREASES ← AND ↓

ELECTRONS ARE LOST MORE READILY ↓ AND ←

SIZES OF THE ATOMS DECREASE → AND ↑

NUMBER OF ELECTRONS IN OUTER SHELL (A-GROUP ELEMENTS) EQUALS THE GROUP NUMBER

Figure 12. Periodic chart of the elements, showing only elements 3 through 86.

pected to be the least active cathode, turns out to be the most active.

In Group VIA all the elements form, with hydrogen, compounds having similar formulas: H_2O (water), H_2S (responsible for the odor of rotten eggs), H_2Se, H_2Te. The boiling points of those compounds increase in the sequence H_2S, H_2Se, H_2Te, H_2O; water is "out of line."

Of course, we chemists understand "why" these anomalies exist; there are quite ready and straightforward explanations. Boiling points depend on the strength with which the molecules attract one another. There are two types of attractive forces, van der Waals and dipole. The former increases regularly from H_2O through H_2S and H_2Se to H_2Te; the second increases regularly in the opposite direction. The combination of those two regular trends results in a superficially more complicated situation (Fig. 13). The case of the battery voltages is analogous, except that the trends of *three* identifiable factors enter into it (Fig. 14). These examples are actually of very simple situations; scientists tangle with problems in which many more factors enter and far less straightforward trends are involved.

All this is intended to illustrate that common sense becomes less reliable a guide, the more developed the scientific specialty becomes. Not that common sense fails to work, but common sense without a knowledge of the state of the art in the discipline can be quite misleading. One can no longer reason from first principles but must begin from what has already been established by earlier workers.

Velikovsky's reasoning is not in itself illogical, but it frequently seems so to a scientist because Velikovsky attempted to use common sense *ab initio;* he started from "first principles" in fields where a whole mass of knowledge had already accumulated over the centuries. That, in a nutshell, is the flaw in *Cosmos without Gravitation* (see Chapter 7): sure, the heavier gases in the atmosphere should settle out—if the only factor involved were the force of gravity; sure, the water droplets in clouds defy gravity—if one does not yet know about colloids and Brownian motion.

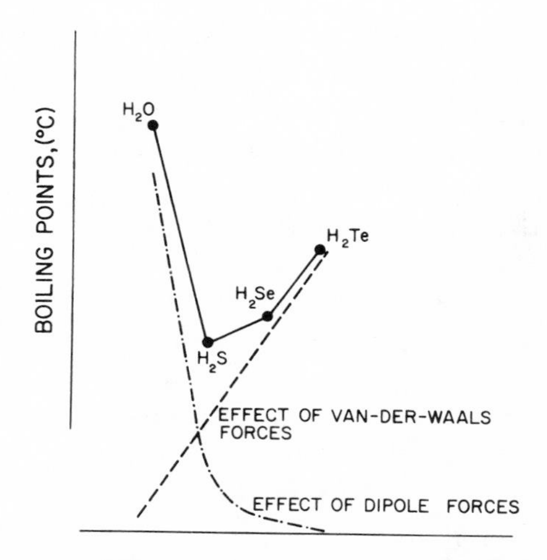

Figure 13. The combined effect of two regular trends is a more complicated sequence.

Too simplistic an approach also characterizes Velikovsky's advance claims. "If Venus has revolved on its orbit for billions of years, there should be no measurable drop in the temperature of the planet that could be detected from its cloud surface . . ." [420]. Fair enough, logical enough—if one assumes that planets were all at one time hot, that their mode of losing heat would enable them to reach equilibrium in a few billion years (is the universe at equilibrium?), if one assumes that the temperature of a planet is a straightforward property rather than one involving the action of many distinguishable factors, if one does not ask how small a rate of change of temperature is in fact measurable.

Again, the idea [408:361–64] that particularly close conjunctions of Mars and earth every fifteen years could be a vestige of contacts between those planets at similar intervals in the past is just too simplistic to have any plausibility. The energies transferred as warring contacters change into serene orbiters would be so great that in such changes any periodicity could well disappear altogether, or the interval of time involved could well change. Only quantitative calculations might serve to justify Velikovsky's speculation; stated qualitatively in this way, it simply carries no conviction at all for a scientist. But to a layman, this could seem to be quite a powerful argument, a common-sense one.

Yet again, the idea that similarities in the periods of rotation and in the inclinations of the axes of earth and Mars could be explained by a past encounter between them; or that Mars is a dead planet because it encountered larger, "more powerful" ones [408:361–64]; or that since methane is associated with petroleum on earth, and Jupiter has methane, therefore Jupiter probably has oil [408:366–71]—all these are statements that inevitably seem fairly logical to a layman but appear to the scientist as simply *non sequitur.* (This is not to say that only scientists can detect the *non sequitur,* but they are used to reasoning about such things, so that even mediocre scientists are able to detect it rather readily.)

This type of oversimplification, making connections on the basis of superficialities, is the same as that which led our forefathers in prescientific days to belief in fortune telling and magic: ". . . if the priest could see some sort of connection or analogy between the omen and the real world, the state or condition of the omen would by analogy foretell the upcoming state of the world. . . . For instance, the reddish color of the planet Mars means to the astrologer that it is magically related with blood, war, and the metal iron, which

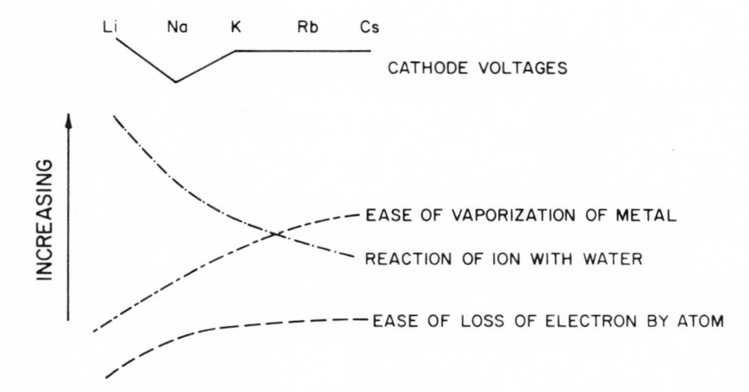

Figure 14. The cathode voltage of lithium is "out of line" because the voltage is the net effect of three trends.

proved so superior to bronze for weapons . . ." [162]. That same sort of thinking—by intuition, poetically, entertaining the possibility of validity in superficial, partly merely semantic, similarities—can be found in Velikovsky's work: "Interplanetary contacts in the celestial sphere are in some respects similar to congress and germination in the biological world. In these contacts the bodies of the planets overflow with lava—fertile ground for vegetation—and comets born of such contacts fly across the solar system and rain gases and stones and possibly also spores, germs, or larvae on planets. Thus the notion of the ancients that love affairs were being carried on among the planetary gods and goddesses is a tale for the common people and a philosophical metaphor for the instructed" [408:361–64]. It is but a short step from that to being prepared to believe that the ancients understood what was really going on, knew about spores and larvae and the rest. This passage may demonstrate a fine poetic imagination, but it has nothing whatsoever to do with science. The mental attitude is anything but a scientific one.

So, common sense is simply not adequate in developed sciences: "many of the results of modern science are difficult to understand. They are far removed from the domain of common sense and everyday experience, and no popularization, however well done, can completely bridge the gap" [28:202]. Perhaps not; but noble efforts have been made in some popularizations. George Gamow wrote some, for instance, *Mr. Tompkins in Wonderland; or Stories of c, G, and h* [111]. In that book Gamow illustrated the significance of those quantities *c, G,* and *h* by describing what one would experience in everyday life if those parameters had very different values. For example, if *c*—the speed of light—were much smaller than it is, we would experience the effect of growing heavier as we traveled at quite moderate speeds, say when bicycling. If *h*—Planck's constant—were very much larger than it is, playing billiards would be an entirely different matter, a gamble rather than a test of skill. I doubt that Velikovsky would have suggested an analogy between atoms and solar systems if he had understood the significance of *h*.

Had he read Gamow's book, he would have had either an inkling of that or at least an awareness that modern natural science deals in phenomena and explanations for which there is no real analogy in everyday experience. Thus electrons in atoms have only a limited number of discrete values of energy available to them. It is as if we could not climb a ramp and were restricted to climbing stairs, moreover, stairs with irregularly spaced steps each of which would need to have just precisely the right height—very peculiar.

So it is not just that everyday common sense is not a reliable guide to speculation in science; even further, there are aspects of science that are plainly contrary to common sense—for instance, being able to climb stairs but not to shuffle up ramps.

The type of procedure used remains the same as a science develops: observational or experimental data are confronted with deductions from theories. But the theories themselves can be, from an everyday viewpoint, paradoxical or simply nonsensical. It is obvious, surely, that nothing can be both a wave and a particle. Yet the only way to deal with photons, electrons, and the like is to use wave equations to describe some aspects of their behavior and particle equations to describe other aspects. How does one know when to use one and when the other? From experience—one gets to know. In other cases it may become even less clear-cut—one has to develop a "feel" for how to handle a particular problem or calculation. Scientists develop a special sort of judgment, by learning the facts and theories in their special fields, until they become adept at fitting and confronting theory and experiment in highly abstruse ways; it is an activity whose particulars—perhaps even whose nature—are similar to what is, from a common-sense viewpoint, a type of insanity, divorced from such human experiences as talking, eating, making love.

Scientists come to accept, without question or with very little question, "theories" that it is difficult not to regard, in the light of common sense, as wrong, even ludicrous. So long as calculations performed in the appropriate manner give conclusions that fit what is observed, most scientists simply do not care what common sense has to say about the basis for the calculations. Dingle's discussion [85:121] is most enlightening in this regard: ". . . physical scientists, almost to a man, have . . . allowed themselves to accept a theory that demands of a clock such an obvious impossibility as that it shall work steadily both faster and slower than an exactly similar one. . . ." He recounts in detail how specialists have told him that this obvious impossibility is not impossible, because their calculations give the right results. That is by no means an isolated case. We are also quite comfortable with the notion that a particle, having insufficient energy to leave a box by scaling its walls (which are impenetrable to particles), can nevertheless "leak through" the walls and thus get out by virtue of its wavelike properties.

Scientific theories, then, are by no means common-sense ones. They can be extremely rigorous in a quantitative sense and at the same time baffling as to what it all "means." But there is always a special sort of reliability about scientific theories: they work. In what the reliability consists, what the relationship actually is between the theory and the reality that it successfully but only partially describes, is a mystery that is debated by philosophers and of no apparent concern to most scientists.

Scientific Judgment

From the beginning of the Velikovsky affair, and particularly after the intervention of the *American Behavioral Scientist,* a major theme was that scientists ought not to have rejected Velikovsky's ideas in such summary fashion, without examining all of them carefully, reading his work "properly," testing his advance claims. For example: "At issue is whether an obviously brilliant and creative scholar who by training is not a 'qualified' practitioner of a particular branch of science (in this case astronomy) can expect to be accorded serious consideration by scientists in that field for a daring but well documented proposal whose acceptance would require sweeping changes in their current thought and outlook" [42].

But in developed sciences (of which astronomy is one) theories can be very abstruse, the ideas involved not being justifiable by common sense or *ab initio* logic. Through familiarity with those theories and with facts and techniques, scientists develop a "feel" for what is likely to be productive. The means by which such judgments are reached, let alone the judgments themselves, cannot be fully appreciated by nonspecialists: "the assessment of plausibility is based on a broad exercise of intuition guided by many subtle indications, and *thus it is altogether undemonstrable. It is tacit*" [321]. Ultimately at issue was a disagreement between scientific specialists on the one hand, and social scientists, humanists, and various interested individuals on the other, solely on a matter of judgment—was Velikovsky's work worth serious consideration by scientists?

Velikovsky's supporters judged, for example, that Velikovsky was brilliant and creative, his proposal "well documented" and of high significance for the sciences. But for scientists, brilliance and creativity have no particular value unless they are coupled with scientific utility; that is a quite basic source of the disagreement. Further, to scientists the proposals were not "well documented"; they were *not documented at all* in the terms of the disciplines involved—physicists and astronomers do not regard folklore and religious documents as constituting reliable data in their fields. So the scientific judgment was based on different premises from those of Velikovsky's supporters, and no resolution was possible: "reasoned discussion breaks down in science between two opinions based on different foundations . . ." [321]. I would add, not only in science.

I am brought to the conclusion, unappealing though it may be, that one has no other option than to respect the specialists' consensual judgment in their field, hoping that unanimity is justified where it prevails and that, where it does not, some specialists will look further into the matter and keep it open. The collective judgment of specialists has been wrong in the past, and will be again, in the case of the really revolutionary discoveries—as was often pointed out during the affair. But there is no possible remedy for that, no means to avoid it (see *Resistance by Scientists to New Discovery* and *Premature Discoveries,* below); there is no substitute for informed judgment, fallible though it may be. The necessary judgment is of plausibility, interest, timeliness, utility—which are not objectively demonstrable—and weight of evidence, its reliability, nature, extent—which might appear to be questions of fact but are actually trans-scientific (p. 313), matters of values and judgment again.

Attempts to substitute the judgment of outsiders for that of the specialists are unwarranted, impertinent even. What is more, if the attempt is successful, the results thereof are undesirable and error rather than truth is served—as in Nazi "science," or in Soviet "science," or where laws are enacted to forbid the teaching of evolution, or to require the teaching of particular ideologies. Even within the disciplines, attempts to impose matters of judgment by fiat are not successful—only the voluntary consensual agreement of most of the specialists can carry the day. That is continually demonstrated, for instance, in matters of units and nomenclature. As sciences develop, nomenclature is often found to become unwieldy, ambiguous, even misleading, as the ideas upon which the names were originally based are modified. Then, periodically, national and international bodies (I am tempted to say, what one must call the scientific establishment) make recommendations for changes to alleviate illogicalities and ambiguities. Most commonly, those recommendations are just so much hot air for any effect that they have; most specialists simply ignore them. If some journal insists on the newly recommended usage, those who are strongly opposed simply find another journal in which to publish. Teachers make up their own individual minds and usually decide to continue with the traditional usage, since acquaintance with it is in any case still needed for the reading of classical articles and books. The traditional nomenclature is replaced only if there is very wide consensus that it ought to be.

Every specialist makes judgments all the time; to what extent to scan the current literature in his field, which and how many articles to read in detail, what research to carry out. Those who make good judgments fairly consistently can later be recognized, for they become the leaders, the authorities. But such recognition comes inevitably by hindsight; there is no other way to reach it with any certainty. Not infrequently, we forecast correctly that a particular individual will succeed or that he will not, but there are as many surprises as correct forecasts, pleasant surprises as well as unpleasant ones.

All of us make judgments, on similarly intangible grounds, as do the experts in their fields. We decide whose advice to take and whose not to take; which books to read, and when to discard them, sometimes after just a few pages or even a few sentences; what is interesting and what is not; what is useful and what is useless; how we shall spend our time. Most of us do not take kindly to suggestions, no matter how well intended or even "well documented," that we change our interests—most particularly not when the advice comes from people who are less conversant than we are with what is involved. I have wondered how de Grazia, for example, would react if a physicist accused him of poor judgment in assessing theories and the value of particular books in de Grazia's own specialty of political science.

Writing in the context of the Velikovsky controversy, a journalist [233] claimed that "if science becomes institutionalized . . . then the right to disagree is unbearably restricted. In a modern technological society science is too important to be left only to the scientists." That is a typical, superficially plausible mixture of sense and nonsense. One needs to make a clear distinction between issues that involve only the scientific specialty itself and issues that are of concern to the larger society. Velikovsky's works have no immediate bearing on social or political policy—they are concerned with fundamentals of geology, physics, astronomy, history; consequently, we must abide by the specialists' judgment. The latter may be wrong (but it is much more likely to be correct than that of the uninformed and uninstructed); if so, it will be corrected sooner or later. In the meantime, no great harm will have been done—countries will not go to war because of it, people will not die or starve therefrom.

Where an immediate public interest is involved, it is a different matter, and then the necessary decision as to values and desirability is rightly made by the wider society. But such decisions have nothing to do with what is plausible, worthwhile, or true from the standpoint of a given intellectual discipline. That official slavery and racial discrimination exist in some parts of the world has no necessary relation to the consensus of specialists in anthropology and psychology about purported racial characteristics and heredity; even in a world free of all racial discrimination those specialists might nevertheless conclude that some groups of people are by heredity different from others. Political intervention may, for a time, determine what specialists say in public and even to some extent what they are predisposed to believe, but in the long run the conventional wisdoms of the disciplines proceed along their own paths. The wider society simply cannot establish anything as worthy of respect in the eyes of the experts if it is the experts' own judgment that it does not warrant respect.

Ethics in Science

Velikovsky's supporters have not, in fact, claimed that the wider society has suffered socially or politically as a result of the rejection by scientists of Velikovsky's ideas. But they have been vociferous about the injustices done to Velikovsky himself and the unfairness of the manner in which he was treated: "If Velikovsky is right, if only in part, then a terrible injustice has been done . . . [requiring] of scientists an act of agonizing reappraisal . . ." [211]; "justice has ultimately to be defined in relation to singular parties"; "is there then no recourse for the scientist who has been damaged by the means detailed in these papers?" [73]. It was argued that, although nonscientists might not be equipped to judge the merits of the substantive issues, they could legitimately have an opinion on the question of fairness: "The layman is really only qualified to comment on the ethical issue: has Dr. Velikovsky received a fair hearing? . . ." [108]. Laymen could naturally have a say on that, since "it would be exceedingly risky to reason that . . . science receives from somewhere a unique moral code that cannot be evaluated by general moral codes" [73]; "the merits of the scientific issue do not alter the deplorable treatment that his ideas received from the profession . . . pre-judging, scorn . . ." [23:78n].

I hope I have made plain already that I hold no brief to defend the specialists' behavior in the Velikovsky affair. Boycott, a role in the dismissal of Atwater and Putnam, *ad hominem* arguments—these are not to be excused; they can be explained, perhaps, but I would not agree to explaining them away. But I must now make the point that there was nothing unfair, no injustice, in the complete dismissal of Velikovsky's *ideas* by scientists, even after only quite cursory acquaintance with them.

First, it is a matter of scientific judgment, how much consideration Velikovsky's ideas (or any other purportedly scientific ideas) warrant—in other words, what a "fair hearing" is. After some 30 years there is still no sign that the scientific community believes Velikovsky's ideas to deserve any attention at all; the original scientific judgment still stands.

Second, how can the merits of the substantive issues be left aside? Had Velikovsky appeared ludicrously wrong, obviously wrong, to the humanists and social scientists as well as to the astronomers, would there still have been an outcry about how his concepts were treated? With how much respect shall we treat the idea that the earth is flat? The suggestion that the substantive issues be left aside is actually made on the basis of a judgment that the ideas have some *prima facie* plausibility, but, for scientists, Velikovsky's ideas have little if any plausibility.

Third, by what moral code shall science be judged? Is there an absolute moral code available from somewhere?

Within the disciplines it is an everyday event that theories are rejected, evidence judged to be unconvincing, proposals for research found not to warrant funding—all on the basis of nothing but the judgment of qualified practitioners in the relevant discipline. Nor is it uncommon for rejections to be given in *ex cathedra* fashion, sometimes accompanied by sarcasm and ridicule, even *ad hominem* remarks. Those who are on the receiving end of such rejections do not always agree that the judgments are correct. They will frequently argue that those were poor, false, wrong, unwarranted judgments, but they do not usually complain that the judgments were unfair or that an injustice has been committed. It is simply accepted that these are matters of *judgment*, inevitably fallible. Poor judgment reflects fallibility, not unfairness or injustice.

Of course, these intradisciplinary rejections are usually private ones. The sarcastic and denigrating comments are known to those who made them, to the editors of journals and the dispensers of funds, and—often in edited form only—to the originator of the proposal. He is not held up to public ridicule and condemnation as Velikovsky was. Are Velikovsky's critics, then, not to be judged for publicly defaming him?

Again, there is something to be said on both sides. Ridicule is certainly unnecessary, and certainly some of Velikovsky's critics could have comported themselves in a more gentlemanly manner. But the public nature of the affair was Velikovsky's doing. It was not his critics who first brought the matter into the glare of wide publicity. They had a choice: to ignore; to answer privately and refrain from public comment, which would have had the same effect as ignoring; or to answer publicly. Yet the ignoring of Velikovsky's ideas by the specialists (for example, by the historians) has also not been palatable to Velikovsky's supporters: they insist that he has, unjustly, not received warranted consideration. But when the specialists do comment publicly, they are told that they should talk in tones of respect about what their judgment leads them to regard as ridiculous. This certainly amounts to being damned if you do and damned if you don't.

Scientists were bound to be criticized for unfairness, injustice, and the rest unless they accepted the judgment of others on matters that are rightfully the scientists' business, matters of scientific judgment. One finds consistently that the accusations of unfairness and injustice were made as a result of disagreement with the *judgment* reached by scientists on the substantive merits of the case.

De Grazia [73] found the scientific community guilty of not adhering to its avowed ideal of objectively assessing ideas and the evidence for them, because that ideal was interpreted by him to mean a right to publish, a right to have one's work read, a right to have one's theories tested. But it would be naive to regard such rights as absolute. Velikovsky, of course, did have the right to publish his work: he could arrange and pay for it himself or he could find someone else to publish it. In the latter case he would need to meet the relevant criteria—in the case of a commercial publisher, the expectation that a profit might accrue from publication; in the case of a scientific journal, that the referees find the material appropriate to the journal and sound as well as interesting in light of the state of the art in the discipline concerned. No scientist, no academic, no individual anywhere has a right to be published on his own terms alone. The scientific journals seek a necessary compromise between rejecting anything that is unorthodox and thereby stultifying progress and, on the other hand, publishing everything sent to them and thereby broadcasting much rubbish and nonsense. So it is inevitably a matter of judgment, what warrants publication in a scientific journal, and only scientists can make that judgment. ". . . Velikovsky was unable to gain access to the accepted scientific journals . . ." [280] ran the frequent complaint. That was Velikovsky's own doing—his work did not meet the disciplinary norms. In the only recorded case of which I have seen any details and where the initial judgment of the referees was that the material might be acceptable, Velikovsky declined to make recommended changes in his manuscript or to argue his case for not making them (pp. 234–35). No practicing scientist expects to be published if he chooses not to meet or argue against the referees' comments.

It is even more ludicrous to suggest that there exists a right to have work read. Manuscripts are read usually by two or three referees and possibly by an editor. Thereafter, if it is published, the work is read by persons who have the requisite interest in the material and by nobody else. How could it be otherwise? And how could one enforce a right to have one's work read? Velikovsky's followers would have scientists read things in which they can find no interest.

A similar objection exists to the supposed right to have one's theories tested. Others will do that only if the theories appear to them to have some interest and some prospective scientific value. I do not expect others to test my theories; I might like it if they do, but I do not regard it as my right. "Velikovsky was unable . . . to persuade those with the necessary scientific resources to perform a few relatively simple tests of his hypotheses" [280] ran the complaint (see also references 172, 433). Scientists and scientific institutions have their own hopes, plans, and programs, and usually manage to keep busy; they are not waiting for suggestions from outsiders for work to do. Why should anyone take time from his own research to do things for which he sees no reason, on the basis of ideas that strike him as quite implausible, when the only conceivable outcome might be some lonely facts, unconnected with the mainstream of the discipline, and therefore of no scientific value? Once again, this complaint comes down to a refusal to accept the judgments of scientists in their own fields.

Now, admittedly, scientists regard the ideal described by de Grazia as a worthy one: "De Grazia was right in contrasting the principles which scientists profess to follow in treating a novel contribution to science with the way they treated Velikovsky's ideas; but these principles must not be applied literally. They should be qualified by their tacit assumptions

. . ." [321]. One of those tacit assumptions is that a judgment has been reached that something plausibly of scientific value is involved.

Scientists do share with other citizens certain commonly held ethical and moral values. But, in the role of scientist, they seek to uphold a tacit professional code of ethics. In that role the highest "ethical" value is ascribed to work that turns out to be right—much can be forgiven if that is the case. On the other hand, something that appears to be wrong is given very short shrift—the originator gets no credit for creativity, originality, brilliance, perservance, honesty, if his idea is wrong; it is simply wrong, period. One is expected to be thoroughgoingly accurate in the reporting of data, the citing of sources, the drawing of inferences. In those areas Velikovsky failed tests that are standard in science: he misrepresented some of the sources that he cited (see Chapter 14, *Word Games*).

It is often said that scientists have an ethical obligation to publish *all* their data, not to select only what supports their hypotheses and to discard the rest. Like other generalizations, this one must be qualified by a tacit assumption. Since all observation and experiment involve "chance" errors of occasionally large magnitude, the discarding of some data is normal, permissible, and expected. The discarded data would naturally be only a small part of all the data obtained; otherwise, possibly, one's hypothesis is wrong and the data correct. Velikovskians have written in outraged tones about radiocarbon tests on articles from Tutankhamon's tomb. Reportedly, the dates obtained were in conflict with conventional chronology and were never published, but would have given some support to Velikovsky's revised chronology (Chapter 5, *Something New, Something Old*).

Now, it would indeed be regarded as scientifically unethical to discard and not report data that would support a hypothesis that is a competing alternative to one's own. But that was not the situation here: Velikovsky's revised chronology does not exist as a hypothesis within history or any other discipline. Those who reportedly obtained these discordant dates did not have a choice to make between two possible chronologies; they had only the one chronology, the conventional one. The dates allegedly obtained varied widely from it, establishing a presumption that something must have been grossly wrong, presumably contamination of the samples (an ever-present possible source of serious error in dating by radiocarbon, giving errors in the direction of more recent dates—which was the case here). Therefore, the data would naturally have been discarded; publication would be judged liable to mislead others into thinking that they were regarded as sound and useful by the scientists who reported them. It may, of course, happen that the dates are later found to have been correct after all, the lonely facts right and the accepted theories wrong; that can always happen, and occasionally has. But the discarding was not an unethical act, as it would have been had Velikovsky's chronology existed as an alternative disciplinary hypothesis. The discarding may later turn out to have been the result of poor judgment but no more than that—not unethical.

It is an accepted point of scientific ethics to give credit, in one's publications, to those who have done important and relevant earlier work on the subject. Once again, judgment is involved as to what is important (which is not necessarily obvious) and what is relevant (which is not always easy to assess). The usual criterion is similar to that in nonscientific disciplines: one seeks to avoid plagiarism, purporting to present as quite new and one's very own something that is only partly new and partly has been anticipated by others. The conscientious assigning of credit is not easy, and it is a rare individual who has not erred through omission in this respect, most commonly through lack of knowledge of the previous work or having forgotten about previous knowledge of it. Such omissions, where clearly unintended, are usually excused, though one apologizes for them and they remain a source of embarrassment. Boring has written [33] about the difficulties of assigning priority, because certain ideas are, so to speak, "in the air" at particular stages in the development of a subject; a Zeitgeist exerts an influence of which individuals are seldom aware. It was the issue of giving Velikovsky credit that sparked the second battle in the Velikovsky affair and led to the involvement of the *American Behavioral Scientist;* it seems in retrospect to have been one of the chief factors in keeping the controversy alive, perhaps even the decisive factor. It was the letter of Bargmann and Motz (Chapter 4) that raised, in the scientific community, the question of assigning credit to Velikovsky for his prediction of radio noises from Jupiter, of the earth's magnetosphere, and of the high temperature of Venus. I shall argue that giving Velikovsky credit, within science, for these predictions is quite unwarranted.

In the first place the degree to which those predictions are correct remains a matter of judgment, and no clear answer is available; at the very least there is room for a genuine difference of opinion. Beyond that, even if the predictions had been accurate and precise, it is no part of the scientific ethic to give credit for predictions made by nonscientific means or in nonscientific work. Thus it was said [15] that Velikovsky deserved no more credit for his successes than Swift for his advance claim, in *Gulliver's Travels,* that Mars has two moons. That comparison was denounced by Velikovskians, but it is apt: *Worlds in Collision* is no more a scientific text or monograph than is *Gulliver's Travels. Worlds in Collision* is admittedly not fiction, but that alone does not make it science; I hope to have shown that its scientific content is misleading or wrong or hand waving. The book is judged by scientists to be not a work of science, and thus no scientist has any obligation to be aware of its existence, let alone to refer to it.

Chemists give little credit to the atomic theory of the Greeks as more than an intellectual curiosity; it was at best a premature idea, of no scientific value, since it could not be used at the time for further scientific work. Nuclear chemists assign no credit to the alchemists for the idea that one element can be transmuted into another; that was a "cheap idea," the alchemists did not adduce evidence to show that it was correct, and in fact their conceptions were at variance with what we now know about conditions under which transmutation can be effected. The technologists of space exploration do not refer in their work to the numerous writers of science fiction who gave plausible accounts of orbiting satellites, voyages to the moon, and the like, even though some of those writings were based on sound scientific postulates and reasoning; again, the value was in the doing, the proving that it could be done, not in the conceiving.

Finally, a scientific prediction is heeded and credited only if the basis for the prediction was a scientifically plausible one: the means of arriving at the prediction must be clearly spelled out, the chain of reasoning displayed. If the latter is found wanting, then the final prediction is accorded no value even if it happens to be correct. Velikovsky's advance claims are scientific *non sequitur:* he did not give clear premises from which the claims logically followed to the satisfaction of those versed in the disciplines concerned. To recall one instance, "in Jupiter and its moons we have a system not unlike the solar family. The planet is cold, yet its gases are in motion. It appears probable . . . that it sends out radio noises as do the sun and the stars" (Chapter 4, *The Advance Claims*). Gases, by definition, are always in motion, but that motion decreases as the gas grows colder; Velikovsky's statement beggars interpretation. On top of that, Velikovsky failed to suggest a mechanism for the emission of the radio signals and did not even specify the type of emission (thermal or nonthermal: in other words, continuous or discrete in terms of the energy emitted as a function of frequency). He did not suggest at what frequencies one might look for the signals, which is crucial. To achieve the necessary sensitivity and ratio of "signal" to "noise," radio-astronomers use apparatus with very narrow ranges of frequency, and to examine a possible source without having any idea of the frequency at which one should look is unlikely to be a productive endeavor. Velikovsky's prediction apparently was based on a superficial analogy between the Jovian and the solar systems, reasoning that is not so much scientific as magical by the principle of correspondence (see pp. 271–72).

So the letter of Bargmann and Motz is an extraordinary anomaly in the scientific literature. It suggests that scientists give credit for predictions based on ideas that the authors of the letters themselves regard as fallacious [22, 260], predictions arrived at by reasoning that the authors themselves regard as invalid, predictions not published in a scientific article or book. Motz suggested that credit was due because Velikovsky was, in his view, a serious investigator and not a charlatan, because of his extraordinary and brilliant mind and his concentrated and penetrating scholarship. But all those qualities are quite beside the point. Scientific credit is assigned according to the criteria discussed above, which pertain strictly to the prediction itself and to the manner in which it was arrived at; any personal characteristics that the predictor may or may not have are irrelevant. Motz himself may have later reached the conclusion that there was no basis on which to assign credit to Velikovsky: in Motz's book *The Universe—Its Beginning and End* [262], published in 1975, there is no mention at all of Velikovsky, despite a bibliography of 30-odd books on astronomy for the general reader, an index that lists the names of upwards of 60 scientists (past and present) whose contributions are noteworthy, and mention within the book of electromagnetic effects and specifically of radio signals from Jupiter.

So the "injustice" done to Velikovsky's ideas was only that scientists judged them to be not science—I repeat, a matter of judgment and not of ethics, morality, law, or justice. The complaints about unfairness in this respect stem from differences of judgment as to whether the work is, or is not, science. No grounds, then, have been established in the Velikovsky affair for a sweeping attack on the ethics of the scientific community in its mode of receiving new ideas into science. The treatment accorded Velikovsky, of course, is a different matter. Finding his ideas to be without scientific merit is no warrant for personal attacks, intellectually dishonest arguing, and the rest.

Interdisciplinary Science

Velikovsky's supporters have ascribed some criticisms of his work to a general tendency on the part of the scientific establishment to work in a hidebound manner within the established disciplines and to denigrate interdisciplinary activity per se [18, 73, 78:1–4, 212, 378, 379, 389, 466]. Once again, I find these criticisms to be partly overdrawn and partly based on misconceptions (though again they contain an element of truth).

What is meant by "discipline" in this context? According to the *Oxford English Dictionary,* it is "a branch of instruction or education; a department of learning or knowledge"; one might also use such terms as "field" or "subject." When one uses "discipline" rather than one of those other words, there is also a connotation of some of the other senses in which "discipline" is used: "schooling"; "the training of scholars . . . to proper and orderly action by instructing and exercising them in the same"; "the orderly conduct and action which result from training"; "a system or method for the maintenance of order; a system of rules for conduct."

Thus chemistry is a discipline that deals with the composition, properties, and reactions of matter. A chemist is one who works in that field in accordance with the accepted rules, and only if he works in that manner—the proper, orderly conduct to which he has been trained. If he works in some other manner, then he is not a chemist—perhaps a chef, or perhaps an alchemist, but not a chemist. So a nuclear chemist or a radiochemist may study the transformation of one element into another through natural or artificial radioactivity or by bombardment with various projectiles (neutrons, protons, etc.). But one who attempts transformation of elements by purely chemical means is an alchemist, not a chemist. The discipline of chemistry teaches alchemy to be impossible, the community of chemists agrees it to be impossible in light of the accretion of experience over many centuries and of satisfactory theory that binds together those experiences into "laws." One who persists in asserting that elements have been transmuted in other ways than those known to be effective would be classed as a crackpot—for instance, one who believes that such transmutations occur in biological systems [188]. Note that this does not *necessarily* mean that the alchemist or crackpot is wrong; it merely means that he is not a chemist and is not practicing chemistry. He is not pursuing work within the discipline of chemistry, even though he may be working with matter and looking for reactions that take place.

One can identify many criteria of this sort to determine whether or not an individual is a practicing member of a particular disciplinary community and can rightly, meaningfully, be called such a member—physicist, historian, even scientist in general. This is not to say that there is a unitary invariant dogmatism within each discipline: there are always disagreements, but these are on matters of detail, and on the frontiers. Consensus reigns on such major issues as types of apparatus or modes of procedure that can give valid data, and theories that are regarded as fundamental, virtually axiomatic for that particular discipline at its particular stage of development. There is something of a consensus also about what degree of probability constitutes acceptable "proof" and what degree of improbability constitutes disproof. Once again, that consensus may not be correct—though it is much more likely to be correct than to be wrong—but one simply cannot ignore the consensus, since it serves to define the discipline.

The existence of disciplines in this sense is a necessary prerequisite to the study of nature in depth and detail. For example, a chemist would get nowhere if he tried to keep his mind constantly open to the possibility that in his chemical reactions there might take place nuclear transformations of some elements into others according to some unknown principle or law of nature. He may be studying the rates at which certain substances react under certain conditions, measuring those rates by monitoring the amounts of the reactants present at various times, *assuming* (until forced to conclude otherwise) that only the reaction that he is studying is actually occurring. The measurements would be useless, meaningless, without that assumption—the data could not be interpreted.

There is considerable pragmatic justification for this mode of procedure: it works. It is used because it has worked, and the assumptions are made because no obvious contradiction of them has been observed. (Of course, at times results do not "make sense," and at intervals such paradoxes lead to revisions of existing theories, assumptions, procedures.) It is important to be clear that there is a defensible, pragmatically based consensus within each discipline, not only about major matters that are accepted but also about which of the presently accepted concepts and theories may need revision at some time and which others will "never" need revision (just so long as the manner in which they are applied is not changed). Ultimately, scientific "laws" are laws because they have successfully tied together, "explained," if you will, actual occurrences. The degree to which laws come to be held as "dogmas" is related to the degree to which those laws have been empirically found to be usable.

Work in a scientific discipline builds on earlier work, and it can do so only by accepting at least some part of what came before. Ideas that remain accepted for a long period of time—for the very reason that contradictions have not cropped up in ever-increasingly frequent instances—are to that extent reliable, and are increasingly relied upon. The "dogma" of science comes after the fact, not before it; its reliability can be demonstrated on request. Do electrons behave both as particles and as waves? We can show you experiments, perform them while you watch, to demonstrate the reliability of that concept. Cases that might seem to contradict this point of view are those that have led in the past to "revolutions" in the mainstream of science—quantum theory and relativity, for example. But in fact those cases merely serve to underline that what is accepted in a given discipline at any given time is not the whole truth. Recognition of that, however, is a very different thing from questioning the truth—in its proper application—of much or all of what is accepted in a given discipline.

It is quite unlikely that something that has proved reliable in the past will suddenly become unreliable. Thus Einstein did not dethrone Newton, nor did he prove Newton's laws to be wrong (nor did he, or any other scientist, ever claim that). Newton's laws are just as valid now as they ever were, within their wide but not infinite limits of application; we now realize that, for the special case where speeds close to that of light are concerned, certain relativistic corrections must be made. Newton's laws remain as a valid *limiting case* of a more general law that applies over a wider range of velocities. Under all ordinary circumstances the corrections to be made for relativistic effects are quite negligible. Thus laws are not replaced by something that is quite different. They are modified as time goes by, usually remaining valid as limiting or particular cases of more general laws.

A discipline, then, is rightly so called because it has built a reliable integrity by fashioning tools, approaches, tests of validity, into something of a self-consistent whole—even though that "whole" is not, and makes no claim to be, any significant part of the totality of human experience or of external reality. Valid and trustworthy interdisciplinary endeavor must respect the individual integrities of the separate disciplines. The most striking examples are perhaps those in which one finds a general thrust by a number of people to use deep knowledge from separate disciplines to illuminate whole new areas. In that way new disciplines are eventually born—biochemistry, pharmacology, physiology, by using varying degrees of chemical insight to illuminate the behavior of living systems; or biometry, a true synthesis of statistics and biology. These new disciplines survived and attained respectability because of their own disciplined character and their lack of conflict with the parent disciplines: they go beyond their parents, but they do not repudiate them. A good biochemist is also a good chemist, and a bad statistician cannot be a good biometrician.

The mere fact that the jargon or subject matter of several disciplines is somehow thrown together does not make an activity interdisciplinary. The synthesis must be carried through with accurate knowledge of the parent disciplines and with respect for the proper use of terms, facts, theories. The very real danger of obtaining nonsensical, misleading results when transferring approaches from one discipline to another has been cogently illustrated by Dingle in his analysis of what belief in the validity of mathematics has wrought—in the way of disaster—when applied to physics [85:122ff.]. I heartily recommend Dingle's book for its sheer interest, its illumination of many facets of scientific activity, and its many priceless anecdotes of scientists in the flesh. Merely to summarize his argument on the point at issue here, and to whet the appetite, I quote Dingle's contention that "the basic misconception of modern mathematical physicists—evident, as I say, not only on this problem [relativity] but conspicuously so throughout the welter of wild speculations concerning cosmology and other departments of physical science—is the idea that everything that is mathematically true must have a physical counterpart . . . the particular physical counterpart that happens to accord with the theory that the mathematician wishes to advocate . . ." [85:124–25].

Another type of interdisciplinary work results when an individual skilled in one discipline turns that skill to work in another area. Linus Pauling, for example, using his broad and deep understanding of chemistry, illuminated features of the structure of proteins that have led to significant new understanding of a disease (sickle-cell anemia) and of evolution in terms of differences in the molecular structures of proteins. Thus Pauling helped found, or put on a firm foundation, the new discipline of molecular biology.

In a third type of example, teams of specialists cooperate to solve particular problems. The best-known case is that of the construction of the atomic bomb, a cooperative effort among mathematicians, chemists, physicists, engineers, soldiers.

Among the examples of great interdisciplinary efforts,

however, one does not find a case in which a single individual has overturned existing accepted doctrines in several different disciplines at the same time, as Velikovsky claims to have done. The likelihood that he is everywhere right—that *anyone* who attempted such a thing could possibly be right—is very small indeed.

There do exist, as Velikovsky and his followers have so often emphasized, barriers between disciplines. Those barriers result inevitably from the success each discipline has enjoyed in refining methods, approaches, and theories: "Today [1976], extreme specialization is a necessity for a physicist who wants to make a meaningful contribution. The different branches of modern physics now speak different languages; each uses its own jargon, unintelligible to those working in other areas" [126]. It is instructive to recall just how narrow a given specialist's actual competence may be, and how uncomfortable he may be in trying to make his own individual assessment even in closely related areas. Some excellent examples are to be found in Dingle's account of his attempts to have answered, by eminent mathematicians and physical scientists, a rather straightforward question concerning the theory of relativity:

> Kathleen Lonsdale [a world-renowned crystallographer] . . . wrote . . . "My difficulty is that I get so far and my mind goes blank. . . . I spent about six months trying to make sense to myself of your paper, but each time . . . my mind just went *blank*. . . . for what my judgment was worth—I could not see the fallacies in it, if there were any. . . . My mind simply does not *care* whether clocks go at the same rate or whether they don't. . . ."
>
> . . . P. M. S. (later Lord) Blackett, who afterwards became President of the Royal Society . . . [wrote] "I am naturally interested in relativity, particularly as I have taught the special theory for many years to our first-year students. . . . However, on the subtle point you are interested in I am afraid I have no contribution to make. I have often consulted my theoretical colleagues on this question, and can find none of them who has serious doubts about the ordinary formulation. I confess I cannot completely follow the details of . . . [your] argument, or

> rather, I have not had the time or inclination to do so. . . . I am no relativist, that is except in the sense that all we practical high energy physicists are, and have not the time nor the ability to discuss fundamentals. . . ." [85:56, 96–98]

This specialization and these barriers are not, as the Velikovskians imply, deliberately erected or maintained by human beings with malice (conscious or unconscious) aforethought. Scientists regret the state of affairs as much as anybody, but it is remediable only to the extent that methods, approaches, and theories can be unified across the borders of disciplines.

One aspect of specialization that is not widely recognized was correctly pointed to by Velikovsky [408:387–89]: specialists in a given field are familiar with its uncertainties, problems, lacunae of understanding, but they often do not realize that similar deficiencies are present in other fields also. So theories accepted from disciplines outside one's own field are sometimes given more credence than they deserve. Thus most scientists accept without question the theory of relativity (see above).

That point does bear remembering when one tries to decide how much credence to give an expert's judgment. I have frequently been delighted at how open and honest a man can be about the limitations and uncertainties of his own work, the tiny area of science over which he has contemporary mastery. He knows the facts and the theories, the problems and the paradoxes, the history of development, and his ego is not threatened as he confesses ignorance and paradox, because he is confessing to mankind's limitations; no other individual gains stature in comparison to himself as he confesses. By pointing to the limitations, he is displaying individual mastery, not individual inadequacy. But take a man just a little out of his field, a little out of his depth, and he is likely to become more dogmatic. He is less certain of himself because he is not familiar with all aspects of the subject, and so he may be even more dogmatic in expounding the certainty of a theory than one who is a master of that theory's special field of applicability. He suppresses doubts because he fears that the true experts might regard him as an ignoramus. Thus Dingle, who thought about relativity for much of his working life, was open about what he saw as paradox or nonsense, but scientists who were not themselves relativists were afraid to give Dingle's discussion any credence, since the orthodox relativists did not agree with Dingle and might label them ignoramuses—impertinent ignoramuses, what is more, trespassing on ground on which they had no right to tread.

Resistance by Scientists to New Discovery

It is widely accepted that science should be open to new discoveries, and that progress in science has been the result of innovations welcomed by scientists and forthwith incorporated into the body of scientific knowledge: "We welcome information that compels modification of the 'laws' we know; we spend our time trying to unearth such information. Nobody is more delighted than the man of science when discordant facts are brought to light . . . ," asserted an astronomer [287]. "The scientist has always been receptive to new ideas . . . ," said another [18]. But the facts are otherwise, and the wish has fathered the thought here.

"Since the turn of the century . . . ," wrote Martin Gardner [112], ". . . widespread opposition by scientists to a legitimate theory, based on verifiable evidence and cogent reasoning, is an increasing rarity. . . ." There is internal evidence here that Gardner was not so sure of the point he was asserting; the statement is qualified to virtual meaninglessness through the use of unquantifiable adjectival judgments: "widespread," "legitimate," "verifiable," "cogent," "increasing." To demonstrate that such "legitimate" new theories are not dismissed, Gardner [113:8–9] pointed out that "Einstein's work on relativity . . . met . . . on the whole an intelligent opposition. With few exceptions, none of . . . [his] reputable opponents dismissed him as a crackpot." Again there are the necessary escape clauses: "on the whole," "with few exceptions," "reputable."

Velikovsky and his supporters may have tried to use the point in *non sequitur* fashion in his defense (the Fulton *non sequitur;* see Chapter 8, *Label with Care*), but they were much more nearly correct about it than were Velikovsky's critics. Larrabee [211], for instance, pointed to the false self-confidence that innovations are nowadays welcomed, and de Grazia [73] emphasized, "The problem that many thought had been solved ages ago—that of recognition of new contributions—turns out to be ominously present. . . ."

Scientists are very rarely exposed, as part of their training, to the history or philosophy of science; when they think about those matters, they readily do so in terms of an ideal, making the naive assumption that actual behavior in science has progressed toward that ideal as the substantive body of scientific knowledge has grown. Most scientists think in terms of the shibboleth of a steady progress to greater knowledge on the part of an open-minded science. The historians and sociologists of science, however, know better and have said so frequently. The evidence seems to be that the genuinely new, revolutionary discoveries are not welcomed or readily accepted. Moreover, that seems to be an inherent and necessary characteristic of scientific activity, not an anomaly, a breaking down of the system, or the doing of dogmatic individuals who suborn the progress of science.

I have pointed out that a scientific discipline is characterized by a consensual acceptance of a body of theory, fact, and methods. That serves to permit extremely sophisticated and deep further probing, experimenting, speculating: by not questioning certain parts of what is accepted, one can investigate dependent parts in great detail. In the terms of Kuhn [204], sets of established paradigms serve as guideposts for normal, mainstream scientific activity. Occasionally a paradigm has to be replaced, and that revolutionary process is accompanied by, among other things, a considerable resistance on the part of scientists to replacing what has served so well in the past. That description certainly seems to fit the classic instances referred to so often in the Velikovsky affair—Galileo, Copernicus, Pasteur, the fall of meteorites—and many others—Helmholtz, Planck, Lister, Mendel.

It is important that one understand the apparent paradox: an orthodoxy is needed to permit orderly, systematic research and progress, yet that very orthodoxy ensures that revolutionary ideas must overcome considerable resistance, thereby apparently impeding progress.

Many have pointed to the useful function served by orthodoxy [52, 321]. In a nutshell, "orthodoxy acts as a kind of filter for new ideas" [67], thereby ensuring that order rather than chaos shall prevail most of the time and that new paradigms, if they are eventually accepted, will have greater utility and reliability than the old ones they replace. A sense of perspective is necessary when confronting an established, reliable paradigm with apparently discordant data. As Boring [34] has put it, it is necessary to suspend "judgment about a small consistency which contradicts a mass of interrelated other consistencies." The expert in any field is familiar with its problems and knows what is possible and what impossible; in consequence he tends to stick rather closely to the tried and true [80]. Nor would we really have it otherwise—my own preference is that my medical advisor prescribe for me what has been known to work in the past rather than use me as a guinea pig. Inevitably, science incorporates new ideas more slowly than, *superficially viewed,* might seem to be necessary, but the slowness is in fact indispensable, lest chaos ensue. Throwing open the doors to all and sundry fascinating new speculations would in fact be retrogressive [125:260].

So, in the normal course of events, science serves to limit or narrow what are seen as possibilities [51:5–6, 373:14]. The common adage that knowledge broadens one's outlook is only one side of the coin; the reverse is that by accepting that some things are true, we thereby also conclude that others are false, and cease to look in those directions for new discoveries—until we are forced at some time to acknowledge the uncertainty of the prevailing paradigmatic knowledge. Normal progress in the mainstream of science does often open up new possibilities for applications, for technology, the building of new gadgets, but it does not broaden the view in the fundamental science itself: that happens dramatically in the revolutionary replacement of paradigms. Orderly work within the paradigms based partly on the achievements of such people as Rutherford, Cockcroft, Walton, Einstein, Hahn, Meitner, Fermi, resulted in applications to atomic and hydrogen bombs, fission reactors, and eventually (in the future) to fusion reactors. At the same time, the very success of the orderly activity in this area ensures that Juergens's and Velikovsky's ideas about nuclear transformations through electrical discharges will receive no hearing at all among practicing scientists in this field.

Conant [55:165, 173] pointed out that "one conceptual scheme may be a block to the acceptance of another," and that a conceptual scheme "is never discarded merely because of a few stubborn facts with which it cannot be reconciled; a conceptual scheme is either modified or replaced by a better one, never abandoned with nothing left to take its place." If for no other reason, science could not accept Velikovsky's notions: he asked that established conceptual schemes be abandoned because of some small consistencies to which he pointed, but he had no new scheme with anything like the necessary overall consistency, scope, and quantitatively verified nature to offer as a substitute.

There exists, then, a useful and necessary scientific orthodoxy, and science resists discoveries that are not consistent with that orthodoxy. Barber [21] cited and analyzed examples to show that resistance by scientists to scientific discovery is a constant phenomenon, and the resistance is greater, the more revolutionary is the discovery. Barber also gave examples to illustrate that resistance is particularly marked when the discovery involves a new method or approach. So de Grazia [73] was whistling in the wind when he said that radicalism in method should not be a deterrent—it is bound to be. Further, resistance runs in inverse proportion to the status of the individual who propounds the new discovery. Scientists inevitably listen with greatest respect to those who have a record of past accomplishment. Linus Pauling, winner of Nobel Prizes, may have met resistance to his proposals regarding orthomolecular psychiatry and the efficacy of large intakes of vitamin C, but the same ideas from a layman would most likely never have been published at all in professional journals. In the furor of the Velikovsky affair, many seemed to forget that Velikovsky had no qualifications at all in astronomy or physics, indeed, had displayed ignorance of some fundamental points. It would therefore have been irresponsible to lend more credence to his statements on those matters than to the statements of Shapley, Payne-Gaposchkin, Struve, Herget, and the other astronomers.

There are sound reasons, then, for scientists to resist new discovery, but "many of those who have noted resistance have been excessively embittered and moralistic. . . . Such bitterness is not tempered by objective understanding of resistance as a constant phenomenon in science, a pattern in which all scientists may sometimes and perhaps often participate, now on the side of the resisters, now on that of the resisted. Instead, such bitterness takes the moralistic view that resistance is due to 'human vanities,' to 'little minds and ignoble minds.' Such views impede the objective analysis that is required" [21]. Moreover, such views fail to take into account the indispensable function served by the orthodoxy that also spawns resistance.

The reassuring thing is that resistance, when "unwarranted," is eventually overcome by the merits of the case. Ultimately science deals in observables, in concrete events, and in the long run theories are accepted or rejected on the basis of their success in explaining data and of the simplicity of their basic concepts [52]. If resistance to a new discovery is not objectively sustainable, the resistance eventually withers away.

Premature Discoveries

Certain ideas, "premature" ones, are simply ignored by science and consigned to a sort of limbo from which they may eventually return; this is not quite rejection of those ideas, though its effect may be similar. According to Stent [380], "A discovery is premature if its implications cannot be connected by a series of simple logical steps to canonical, or generally accepted, knowledge." As an example, "Avery's discovery [that DNA is the material substance of which genes are composed] was premature . . . not appreciated in its day. . . . I do not mean that Avery's discovery went unnoticed, or even that it was not considered important. What I do mean is that geneticists did not seem to be able to do much with it or build on it. . . ."

Other examples of prematurity have been given by Bernstein [28]. Experiments performed in 1928–30 showed that parity is not conserved, something that was "discovered" in 1956:

> at this time [1928–30] the study of weak interactions was in its infancy, and there was just no theoretical context in which to put the results. . . . They were . . . a kind of statement made in a void. It took almost thirty years . . . to enable physicists to appreciate exactly what those early experiments implied. . . . in 1929 . . . H. Weyl . . . noticed that a remarkably simple and beautiful mathematical theory . . . was available, but only if parity conservation was abandoned. For many years, the theory was ignored. After the developments of 1956, this theory was revived . . . and has become the commonly accepted mathematical description of the neutrino. [28:65–66]

It is inherent in Stent's definition that a discovery is premature when it cannot be used in the mainstream of science. That certainly is an apt way to class Velikovsky's ideas—they cannot now be connected to the body of existing canonical knowledge. Therefore, quite irrespective of the "truth" of the ideas, they could not be taken up by scientists. This phenomenon of prematurity can often explain why subjects remain on the fringes of, or outside, science; for instance, "in the absence of a hypothesis of how ESP could work it is not possible to decide whether any set of relevant observations can be accounted for only by ESP to the exclusion of alternative explanations" [380]. It can be a useful exercise, when confronted with some startling new suggestion, to ask how the implications of that suggestion could be used in the mainstream of scientific activity. If not—as in the case of

Velikovsky—then science will have nothing to do with the idea. It may in fact be nonsense, or it may simply be premature.

Scientific Activity

Nonscientists often seem to think of science as a coherent monolithic structure; in a sense, that is already implied in the very use of such comprehensive terms as "science," "scientific," "the scientific establishment." The only factor that seems to characterize science as a whole is that, in the long run (which may be very long indeed), untruths are weeded out and what remains becomes more reliable. In all other respects there is diversity: strictly controlled experimentation but also pure observation with very limited or no effective control over the phenomena studied, inductive construction of theories from facts as well as purely deductive reasoning, common sense as well as a very special kind of logic that can easily look like nonsense, acclaim of novelty and a search for it as well as inherent resistance to novelty.

The last statement can be refined. I have discussed the inaccuracy of the notion that science readily embraces new ideas. Resistance to the new is evident when the new actually contradicts an established paradigm, when it is *too* new. But much that is "new" is relatively, even quite, consonant with the conventional wisdom, and that type of novelty is indeed continually sought and quickly acclaimed and accepted. So, for example, once understanding of the structure of atoms made it possible to see how one element could be transformed into another, transmutation ceased to be an alchemical fallacy and became a scientific concept within an established paradigm; increasing numbers of scientists sought new reactions that would lead to new types of atoms—novelties within the conventional. There are innumerable possibilities in every field for discoveries that amplify existing knowledge, and the overwhelming majority of "pure" scientists is preoccupied with pursuing those possibilities, seeking to make foreseeable discoveries. The revolutionary discoveries inevitably come also, but most commonly as by-products of conventionally oriented work; the revolutions are unforeseen because they are unforeseeable.

Large numbers of scientists are at work in innumerable specialties and subspecialties, often with little communication across the borders. Each group has its conventional wisdom, its jargon, special techniques, laws, even theories. I think the best short description of this reality is Polanyi's phrase, the "republic of science" [320]. Science is pluralistic. It is an activity characterized by the opportunity to make discoveries. The laws that result are determined by the nature of the task, which is simply to make discoveries—and that description is accurate even though it may appear to be circular, tautological, or paradoxical. The overall task is indeterminate because there is no final goal in sight, but in working at it there has come to exist a definite structure. In science there is "self-coordination by mutual adjustment"—tacit agreements and assumptions are shared and individuals voluntarily (partly as a matter of personal self-interest) accept the consensus. There is "discipline under mutual authority"—there is no permanent set of accepted leaders, authorities, or an "establishment"; the authority resides in the tacit "constitution" of the republic.

To de Grazia's questions [71]—"Who determines scientific truth? What is their warrant? How do they do so?"—one has to answer: scientific truth is determined not by identifiable individuals but by consensus; the warrant is not anyone's to give or to withhold; in the long run the unreliable is weeded out, and scientific truth is self-determining, emerging from disciplined activity according to tacitly accepted rules, after much trial and error, the only criterion over the long haul being reliability. The character of science is determined by the task of science, and is influenced only temporarily, briefly, occasionally (and usually quite ineffectively) by any who seek deliberately to design a structure within which scientific activity shall take place according to some formula. De Grazia's notion [73] that one can pass judgment on science and reform doctrines, formulas, and tactics is naive; all that one can accomplish by fiat is to impede science, the process of discovery.

Most scientific activity is carried on at a quite mundane level, by rote and routine in the acquiring of semi-predictable and not-too-startling novelty. The marvels of science, the periodic revolutions that may seem to the outsider to characterize science, are but one aspect of it—an important but very small part in terms of the total effort spent in all scientific activity.

Science proceeds unself-consciously. The philosophers discuss epistemology, but to scientists that is an irrelevancy. Scientists uncover things that work and that can be put to use (to *scientific* use), and those are the justification for what they do; no other is needed.

Science is a pluralistic activity, with inconsistencies, paradox, triumphs, and failures, periods of rapid development and periods of stagnation. In tones of indignation Cantril [46] concluded that "in theory, science's 'reception system' is based on the objective, rational evaluation of submitted evidence. In practice, however, . . . it operates much like other social institutions, complete with hierarchy, dogma and coercive power. . . ." But of course—how could it be otherwise? Science is an activity carried on by human beings, so all those elements are inherent in it. But there is democracy as well as hierarchy, openness as well as dogma, lack of coercion as well as coercion by tacit assumptions: ". . . science is a human activity and . . . you have, therefore, to take into account the properties of human beings when you are assessing facts and theories . . . " [33] or any other facet of science.

Science as an Institution

Since science is unfamiliar territory to most people, it has an air of mystique about it; it is mysteriously powerful, mysteriously correct, populated by beings identifiably different from other human beings. Given the mystique, misconceptions are inevitable and resentment of science and of scientists endemic. Perhaps some of that mystique can be stripped away without going into details about the subject matter of science at all. Given that the nature of scientific activity is determined in part by the fact that it is a human activity, an outsider's view of science might be quite realistic if he saw it predominantly in that light, and applied to it his understanding of other human institutions. There are many analogies.

The army is also a human institution. It also serves a necessary function within the wider society, interacts with other institutions, is supported to an extent that is determined by the democratic political process with its pressure groups and lobbies pushing every which way. But the army also has autonomous control over much that is regarded as its own business: it makes judgments about training recruits and selecting for promotion; about what has military value and what does not; about strategy and tactics; about what is efficient and what is not; about what are effective approaches to problems and what are not; about which new developments are worth pursuing and which are crackpot ideas.

Like science, the army exists in its present form because, by and large, that has been found to work. Like science, the army is fallible, and that fallibility is of no concern to the wider society most of the time—only in the intermittent crises of wars does the public take much notice. Again, in these crises outsiders sometimes seek to substitute their judgment for that of the specialists; similarly, though the outsiders' judgment may occasionally be correct, in the long run it is less likely to be so, and in matters that are purely military, as opposed to transmilitary (see p. 313), one had better follow the advice of the specialists and respect their judgment.

Most of the time the army—like science—is wondrously powerful and effective. But occasionally it comes up against something quite new, and falters. What works in regular warfare breaks down in guerrilla warfare—as the Germans found in Yugoslavia, the French in Algeria and Vietnam, and the Americans too in Vietnam. Then there emerge the wise outsiders who call for reform, to make the institution really effective in such a new type of situation, forgetting that the main task remains the regular one, and that one cannot design an institution to be at the same time orderly and efficient in dealing with regular and well-understood problems and also effective with idiosyncratic problems. Hard cases make bad laws.

The military has its disciplined procedures so that order rather than chaos will prevail, and most of the time those procedures work. The guidelines ensure that any human being of average intelligence and competence can adequately perform the necessary tasks. Similarly, the tacit guidelines and methods of training recruits to science ensure that any human being of average intelligence and competence can carry out scientific tasks in a workmanlike fashion. Everyday science, like everyday military activity, is mundane routine, and most of the participants in it nowadays are rather average people. Science, like the military, has its hordes of privates and non-coms, as well as its few heroes (from all ranks) and its few field marshals.

The military has its code of justice, which can, of course, be evaluated on the basis of other moral codes. But it is a type of justice suited to the overall needs of that particular institution, and what appears to outsiders to be rank injustice and unfairness may nevertheless be a practical necessity for the institution, and thereby for the wider society as well.

In the army, as well as in science, there is freedom within the bounds of discipline and responsibility. There are irksome duties as well as positions with prized perquisites. In theory, anyone can reach the top; in practice, few are capable of it. Those who do reach the top display a paradoxical mixture of conventionality and unconventionality, of creativity within prescribed limits.

Like science, the army has a set of ideals to which it strives to conform. Surprisingly, as with science, the performance does not fall all that far short of the goal, even though many individual practitioners deviate considerably from the ideal, and few exemplify in themselves all the desired characteristics. The performance of the institution transcends the limitations of the individuals who comprise it, as a result of the discipline, task, code of justice.

The institution of science, then, differs from other human institutions in those details that pertain to its specific task and subjects of concern but not in any other way.

Scientism

That science is synonymous with certainty and truth is an insupportable belief but, like many other insupportable beliefs, it is held and acted upon by a significant number of people. That specific belief is often described as "scientism" or as reflecting "scientistic attitudes." I gave some examples earlier (Chapter 13, *Tit for Tat*) of statements that betrayed scientism on the part of some of the actors in the Velikovsky affair; similar examples are available in profusion in our culture: advertisements that stress that tests or proofs are "scientific," for example. It is only through this usage of the words that scientistic beliefs are clearly revealed, because very few people now do not realize—once they think about the matter explicitly—that scientism is an intellectually insupportable attitude.

We all have the ability to hold unconsciously to beliefs that we recognize intellectually to be invalid; unless we are extraordinarily careful, we frequently act out of our unconscious beliefs rather than from our best intellectual insights. In Chapter 8 I showed that effect in the actions of crank labelers: they recognize intellectually that one can rarely if ever be justified in calling someone a crackpot, yet their convictions are so strong in particular cases that they do so label people. In Chapter 9 I pointed to this tendency displayed by Velikovsky, in Chapter 12 by the Velikovskians: the general intellectual insight that no individual, Velikovsky included, could always be right was submerged, in the heat of the controversy, by the thobbing conviction that Velikovsky had not in fact been wrong on any significant point. That same phenomenon causes scientists all too often to speak and act as though science were certain and true even though they recognize intellectually that this is not the case.

That was perhaps the greatest failing of Velikovsky's critics: they were not willing to make their case the demonstrable one—that Velikovsky had not proved his scenario, that his claims about physics and astronomy and science generally are very implausible. The critics wanted to establish that Velikovsky was wrong without any doubt, and they sought to invoke the authority of science on their side. Carl Sagan illustrates the phenomenon nicely. He clearly recognizes that the argument from authority is inadmissible, and takes commendable public stands on that in general: supporting AAAS symposia about UFOs and about Velikovsky, seeking to enhance public understanding of science by writing for a general audience, refusing to endorse an *ex cathedra* statement about astrology [355]. Yet in particular cases, notably that of Velikovsky, his utter conviction on the specific issue leads him to try to obtain assent to his opinion even though he cannot prove his opinion to be correct, and to invoke at least implicitly his standing as a scientist in seeking assent.

So Sagan set out to calculate the probability of the series of cosmic collisions postulated by Velilkovsky. He erred, or misrepresented Velikovsky, in treating the series as a set of independent events [160]; he used distances that describe actual physical contact, though Velikovsky never specified such contact and a reasonable interpretation of Velikovsky's writings is that only relatively close approaches were what Velikovsky had in mind [100, 192]; and he concluded, "Hypotheses with such small odds in their favor are usually said to be untenable . . ." [353, 356–59, 361]. That last statement illustrates my main point here: Sagan seeks to use the statistics not merely to establish "implausible" but to convince of untenability. Yet he must know that any event of finite probability can occur and that Velikovsky never claimed that his scenario was probable, only that it happened. Sagan's appeal to probabilities is quite beside the point, since he seeks to refute Velikovsky beyond doubt [192, 259].

Scientistic attitudes and statements bring science and its practitioners into ill repute, and do not serve the causes sought by those scientists who engage in public discussions. Unfortunately, the lesson has not yet been widely learned. In the debate over creationism and evolution in school curricula, for example, some well-known scientists and science writers have insisted publicly that evolution is not a theory but a scientific fact [70:292, 129,196, 216, 360:27].

In the Velikovsky affair quite valid criticisms were made by Velikovsky and his supporters of the scientism underlying the behavior of the most intemperate of his opponents. I have given my view that those criticisms, by de Grazia in particular (Chapter 4, *Something Amiss in Science;* Chapter 12, *Antiscience;* Chapter 15, *Ethics in Science*), were too extreme in castigating science as a whole and scientists as a class for the actions of a few astronomers and others. De Grazia is correct, though, in pointing to the *tendency* for scientists to be scientistic and not deeply understanding of the social sciences and humanities even when those disciplines are relevant to the scientific enterprise—as in the making of science policy, the educating of students not specifically studying science or technology, the broadening of curricula for students of science and engineering. Recognition that such problems exist has become more general in the last two decades, leading, for example, to the mushrooming of courses and programs called something like "science, technology, and society" [476]. But we are far from eradicating scientism. It is not only scientists, or some scientists, who find it difficult to discard the unconsciously held belief that science offers certainty and truth: our whole society looks to science as the provider of certain answers. Even those who attack science do not question its power to establish the truth: they ask that science be reformed, they complain that science does A instead of B, but they also believe that, in principle, science offers truth and power over nature. Only when the limitations inherent in science become generally appreciated will the belief of scientism begin to relax its hold on us all.

16

Analogous Cases

> Though analogy is often misleading, it is the least misleading thing we have.
>
> —Samuel Butler

Is the Velikovsky affair unique, a singular occurrence whose like was not seen before or will be seen again? Or does it typify some larger class of situations?

The second is closer to the truth, it seems to me. The controversy is an example of "the kind of confusion that must be expected when technical matters become the focus of public debate" [236]; it "raises what is perhaps the most disturbing question about the public impact of science. How is a layman to judge rival claims of supposed experts?" [127]. The problem arises, by definition, when there is substantial division of opinion among the "experts," and that very division is *prima facie* evidence that no clear-cut factual, technical resolution of the issue is available. One of the prime lessons to be learned from the Velikovsky affair is that specialists are prone to maintain that clear technical answers are available when in fact they are not. Like all of us, the specialists form an opinion as to the probability that something is so, and thereupon speak of what seems a high probability as a certainty.

In any public controversy about science or technology the technical experts must recognize that "in a democratic society, the public's right of access to the debate . . . is as great as the public demands it to be . . ." [455]. In the Velikovsky affair scientists chafed when they found their authoritative judgments not being accepted without question by the public (or by intellectuals and scholars who were not scientists). Many experts will have sympathized with the astronomer who thought that the "rare near-unanimity among scientists should have completely disposed of the question" [247]. The affair will have been of considerable value if scientists learn from it that it is not so easy to win the day in public debate, and if they further learn that their opinion is not fact. In such public controversies "the most science can do is to inject some intellectual discipline . . ." [455] into the debate.

Scientists tried to do much more than that. They sought to demolish Velikovsky's case and, by blundering in the attempt, in the end accomplished far less. Perhaps, as Damon Knight suggested, Velikovsky's critics were careless because they were debating with nonscientists rather than with their peers and in the accustomed manner in their own discipline: "it did not seem to matter . . . whether the arguments used were correct or not, since it was known that Velikovsky was wrong—if not for these reasons, then for other reasons. . . . The same assumption also led Velikovsky's attackers, by a familiar psychological process, to assume that since Velikovsky was wrong, *evidence to that effect must exist* . . ." [191:136–37].

Perhaps that does explain the lack of care and rigor shown by Payne-Gaposchkin [287, 289] and by Menzel [247], for example. At any rate, the lessons are clear enough: in public debate scientists had better limit their authoritative statements to details where their competence suffices to carry the authority; they had better state clearly that unanswerable questions are indeed unanswerable; they ought to recognize that their greatest contribution can be to inject intellectual discipline into the controversy. If these lessons are not heeded, then the results may well be quite unpleasant. According to Weinberg [455], "To a scientist, Velikovsky is not to be taken seriously because he did not conform to the rules of procedure of the republic of science; to the public, he is the victim of an arrogant elite. The nonscientific public came close . . . to demanding the right to pass judgment on scientific questions." Such an unpalatable fruit of public argument would be made no tastier for scientists by the fact that they themselves had sown some of the seeds.

Many scientists feel the urge, responsibility, duty, to set the public straight. They recognize the lack of public understanding of what science is all about, and talk about ways of changing that state of affairs:

> . . . scientists and their official representatives must take the initiative in combating pseudo science—and, on the positive side, presenting science and the scientific method in the best possible light. . . . An American who has not been trained in science may sincerely feel that Galileo and Velikovsky are both martyrs of "authoritarian" tendencies. . . . What shall we tell him, and how shall we put it? . . . Science has a message beyond the results it has achieved in medicine and technology. It has not yet succeeded in communicating this message to the American public. . . . [317]

> That there is a great need for . . . scientific books directed to the general reader was apparent to everyone. . . . the publishers frequently print, . . . and the public likes, . . . the *wrong* kinds of scientific books . . . *Worlds in Collision* or *Dianetics.* More sound, more responsible, though less sensational works, even when written with an eye to the general public as audience, are seldom as widely read. . . . [188] (See also references 207, 268.)

So far, so good; the problem is recognized, and there is a desire to do something about it. Yet a number of critics argued, or explained, that the task of effectively countering Velikovsky was too time-consuming to be worthwhile [37, 238, 288] (Chapter 3, *The Case for Velikovsky*); it was also said (and by some of the same people) that it was important that the job be done. If it is indeed important, those who think so need to do the job; they cannot expect others to do it for them. Further, if it is really important to educate the public generally as to what science is about, are not Velikovsky and his ilk heaven-sent opportunities, to be used as illustrations of what science is and what it is not? After all, the public interest is there to be capitalized upon. As more and more matters involving science become public concerns, scientists had better learn that they must *compete* for acceptance of their viewpoints. If they wish to be believed, they must find more persuasive arguments than their opponents (bearing in mind who it is that they wish to persuade—not, usually, other scientists). Velikovsky's critics, on the whole, did not compete effectively.

There were many good reasons for holding Velikovsky's scenario to be highly improbable, but the astronomers and physicists characterized it as "impossible" on grounds of celestial mechanics, which is simply not so (Chapter 5, *Scholarly Support*). The case of the Loch Ness monsters is similar in this respect—it is staggeringly implausible that these animals should exist, but not totally impossible; almost all the experts (zoologists in particular), however, have stated it to be impossible.

Within every discipline there exists something of a consensus as to what is fact, what are the reliable theories, and what types of changes are likely to be needed in the future in those theories. Within that broad agreement, many points of detail are continually argued. Such arguments are usually resolvable because of the prevailing consensus, for instance, as to what constitutes proof—what level of probability can safely be translated into certainty. These are more or less regular and orderly arguments within the disciplines; they are continually being resolved and new ones begun. This consensus prevailing in each discipline has been aptly called by Polanyi [321] a tacit one; it is not usually made explicit. As a result, most scientists in the normal course of events do not live with an awareness that they are dealing with matters of high probability—they think in terms of certainties, since their colleagues and peers believe as they do. Scientists do not habitually emphasize that (their) belief is one thing and truth another. Not only are scientists certain about their facts, a certainty that is at least pragmatically justifiable; they are also certain that the theories that *work* are *true.* Philosophers recognize that the truth of a theory is much less certain than that of the facts, and that theories could be more justifiably prized solely on heuristic grounds. Worst of all, scientists come to believe that their judgments (collective, if not individual) on all discipline-related matters are right—for which belief there is much less justification.

These attitudes of certainty are inevitably communicated to students and to the public, who then quite naturally expect to be able to get definite answers from the scientific community about anything that is related to science. All too often, scientists respond with definite answers. Such answers, however, are inappropriate to many types of questions.

With Velikovsky and Loch Ness, for example, a scientist might well hesitate to be certain, because it is so hard to prove—perhaps, indeed, impossible—that such a thing is definitely not so; the difficulty of establishing an *absolute negative* is virtually a cliché for philosophers. That is highly germane to present concerns about environmental hazards, for instance. Weinberg [455] has defined the category of trans-scientific questions: "epistemologically speaking, questions of fact and can be stated in the language of science, [but] they are unanswerable by science. . . ." Weinberg gave a number of examples, of which I shall mention two.

What effects might result from the exposure of humans to very low levels of radiation over long periods of time? One suspects that mutations will occur, and that cancers might be initiated, for example. Weinberg points out that one could, in principle, have tested the effects of the standard exposure that was decided by the Atomic Energy Commission to be permissible. However, to find evidence of 0.5 percent increase in the rate of mutation, one would need to expose 8 billion mice to obtain a statistical level of confidence of 95 percent (in other words, there would be 1 chance in 20 that the answer obtained would be wrong). That experiment would be very expensive and time-consuming, and the answer would not be certain—only 95 percent certain, minus the uncertainty incurred in taking mice as good models (for this purpose) of human beings. To achieve greater certainty, or to detect increases of less than 0.5 percent in the rate of mutation, would require correspondingly more unmanageable experiments. The important lesson is that "one can never, with any finite experiment, prove that any environmental factor is totally harmless. This elementary point has unfortunately been lost in much of the public discussion of environmental hazards."

The actual question of public concern is, how much exposure are we going to tolerate, given that risks are involved whose magnitudes cannot be determined with exactness? Various experts will give different estimates of the likely number of additional deaths per year, for example, but those numbers are highly uncertain. What has to be weighed is a *range* of numbers of possible deaths as against such possible or likely other undesirables as a shortage of energy and corollary unemployment. Even if there were exact numbers available, no scientist, physician, economist, or other expert would be qualified to give an answer. We have to answer on the basis of values and probabilities, not on the basis of facts.

Another example of a trans-scientific question is the probability of such an extremely improbable event as a catastrophic accident involving a nuclear reactor, or an earthquake powerful enough to destroy the Hoover Dam (or, I would add, Velikovsky's scenario of cosmic events, whose probability several scientists attempted to calculate [289, 305, 385]). Weinberg points out that "probabilities of such events are sometimes calculated. . . . But the calculations are obviously suspect . . . because there is no proof that every conceivable [factor] . . . has been identified . . . [and] because the probability is so small [that] . . . there is no practical possibility of determining [it] . . . directly. . . ."

Herein, then, lies a seed of the public controversies: trans-scientific questions are posed to scientists, who as a matter of habit give a judgment, an opinion, in the tones of certitude. If the answer is palatable to the public, perhaps not much harm is done—the same conclusion would have been reached through a consensus about values. But if the answer is unpalatable, the fat is in the fire. As Polanyi [321] has put it, "Laymen normally accept the teachings of science not because they share its conception of reality, but because they submit to the authority of science. Hence, if they ever venture seriously to dissent from scientific opinion, a regular argument may not prove feasible. It will almost certainly prove impracticable when the question at issue is whether a certain set of evidence is to be taken seriously or not. . . ."

There we have another piece of the explanation of the Velikovsky controversy: it was messy, confused, not a "regular argument," because there was no substantial area of tacit agreement between the arguing parties. If public controversies are to be meaningful, informative, useful, then it is vital that common ground of some sort be established. The necessary common ground might in fact be no more than an acknowledgment that the question (or some part of it) is indeed trans-scientific and not purely technical. But agreement on what might seem to be so obvious a point is not easily arrived at. The specialists are reluctant to admit their own limitations and those of their specialty, particularly since they are not necessarily clear themselves about those limitations. Weinberg has pointed to the difficulty of achieving the necessary common ground in these arguments: "in trans-science . . . matters of opinion, not fact, are the issue. . . . One must establish what the limits of scientific fact really are, where science ends and trans-science begins. This often requires the kind of selfless honesty which a scientist or engineer with a position or status to maintain finds hard to exercise. For example, in the acrimonious debate over low-level radiation . . . neither side was quite willing to say that the question was simply unresolvable, that this was really a trans-scientific question . . ." [455].

Because of the public importance of many trans-scientific questions, ways have been proposed of clarifying arguments of the sort that have raged in the media and in public hearings of congressional committees. One proposal is that a Science Court be established, in which the disagreeing experts would dispute the technical issues before a panel of technically trained judges. The rationale for a Science Court has been put [364] in this way: "In many of the technical controversies that are conducted in public, technical claims are made but not challenged or answered directly. Instead, the opponents make other technical claims, and the escalating process generates enormous confusion in the minds of the public. One purpose of the Science Court is to create a situation in which the adversaries direct their best arguments at each other and at a panel of sophisticated scientific judges rather than at the general public. . . ." So the Science Court would serve to separate opinion from fact and produce an agreed set of facts to be presented to Congress and to the president: the uncertainties spelled out, the ranges of probability of risks evaluated, but no answer attempted to the question, "should this be done?" That last, trans-scientific question is one of values, to be settled by the political process and not by the specialists.

Such proposals are intended to establish a proper role for the experts: to give the best available technical advice but not to couple that with opinions about desirability. Scientists have not filled that role satisfactorily in the past, neither in the Velikovsky affair nor in other arguments. In the dispute over antimissile missiles an observer was moved to make the plaintive query, "If scientists do not scrupulously guard a certain minimum of detachment and self-restraint, what do they have to offer that the next man does not?" [455].

An awareness of what experts can legitimately say, and what their proper role is, would have been valuable to observers of the Velikovsky affair. It would have been even more helpful if the experts had restricted themselves to such statements and to such a role. We can learn from the Velikovsky affair that in other controversies also we are likely to get more from the specialists than they really have to offer—that is to say, more advice but less help. Technical experts, after all, are human beings as well as specialists and hold beliefs about all sorts of political, religious, and social matters. As with the rest of us, their views on particular issues may be colored, even predetermined, by ideological considerations: "on any significant public issue—scholars will . . . be found to support both sides. And faced with that choice of expert opinion, it is . . . not easy . . . to resist the temptation to believe that the scholar who shares [one's] . . . approach is the man with the objective facts" [236].

All that seems obvious once it is said, but it is nevertheless easy to fall into the trap. Not many years after World War II there was a bitter debate whether or not to proceed on a "crash" program to develop a hydrogen bomb. The main technical issues were: whether a hydrogen bomb was feasible at all (a different principle was involved from the Hiroshima-Nagasaki weapons: nuclear fusion in contrast to nuclear fission), and whether a crash program would be more efficient, or even necessarily speedier (because of the absence of fundamental scientific understanding), than a more normal program of development. The best-remembered spokesman in favor was Edward Teller, opposed, Robert Oppenheimer. Most disengaged observers have seen no reason to doubt that both these men were consciously moved only by sincere and honest conviction, but unconscious motives are something else again. Teller, Hungarian emigré and vociferous anti-Communist, believed and said that the required technical knowledge was already available and that a crash program would succeed. Oppenheimer, a liberal who had dallied further on the political left, believed and said that the required technical knowledge was not yet available and that a crash problem was therefore not warranted.[1] Would anyone care to argue that it was mere coincidence that these men reached technical conclusions that were consonant, in terms of the policy indicated by those conclusions, with their political views, with their assessments of the immediacy and extent of the threat posed to the United States by the Soviet Union? And even if that was coincidence, it was surely not coincidental that we laymen

1. For decades thereafter there was still no general agreement about how adequate the technical knowledge then was. Documents made public in 1982 indicate that Oppenheimer was correct on the technical question [36].

aligned ourselves with those experts along political lines: the liberals believed that Oppenheimer was right, and the conservatives were convinced that Teller was correct. That was not an intelligent way to go—to believe an expert, on a matter of nuclear physics, because his political views happen to be congenial. How to avoid this trap?

The important thing is to be clear about what the central point at issue actually is. Where eminent men disagree violently, and both sides present their cases as proven, we can be rather sure that certainty is not in fact available, and that the matter is not technical but rather trans-scientific. It is a dispute over probabilities, values, desirability, *not* over facts. One must then seek to disentangle the arguments in that light, bearing in mind what can be learned from, for example, the Velikovsky affair: that the debaters (as we do) have a vested interest in believing something—religious, political, whatever—and inevitably derive a certain personal satisfaction as a result of those beliefs; that expressed arguments will therefore be replete with wishful thinking, uncritical commentary, preconceived notions about ancillary matters; and that words and numbers will be used in tendentious and misleading ways. Beyond that, one does well to keep in mind at all times that the experts are human—they vary in competence in their fields of expertise, and there are in fact always a number of quite technically incompetent "experts." Even granted competence, they vary in general intelligence, in honesty, in the ability to say what they mean, and in all other human characteristics.

Epilogue

I have thought a great deal about the Velikovsky affair, and now I have written quite a lot about it. Most of the time I thoroughly enjoyed what I was doing, aware that I was learning much in the process. I have not, however, learned a simple answer to the question that is usually put—was Velikovsky right?

I did learn that Velikovsky was quite ignorant of science, and I cheerfully dismiss as nonsense his *explanation* of physical events, where indeed he gives any. Nonetheless, his cosmic scenario just might have some truth to it; I severely doubt it, but I cannot disprove it. There does seem to be something unsatisfactory about the conventional chronology of Egypt (the Sothic periods, for example), and perhaps some of Velikovsky's revisions will turn out to be a better approach to the truth. But I claim no certain global answer.

Despite that, I think the quest has been highly successful. Most narrowly, I have the satisfaction of having become an expert; when I now read some new piece of writing about the Velikovsky affair, I am usually able to spot errors and misinterpretations and red herrings very efficiently, and that is a source of personal satisfaction. Going a bit further afield, I find that my reading on other subjects (most especially on the fringes of science) is more critical and analytical; neither advocates pro nor advocates con have as easy a time with me as they used to. Beyond that, I find that I am better able to discern the important points of the seemingly technical matters of which the media are full—the controversies over nuclear power, the ozone layer, Laetrile, saccharin, and the rest.

I would be very pleased if the reading of this book were to be of like help to others. I should like my readers not just to accept my analysis and interpretations but to be stimulated to do as I did—to go to the original material and work it out for themselves: not necessarily with the Velikovsky affair (though I would find some satisfaction if I had stimulated that), but certainly with the many issues of social importance that concern us all. Our right to vote is of immeasurable value, but that value is realized only in proportion to how well informed and critically minded we are as voters.

Faced with a question that concerns us, the only rewarding way to seek an answer is to proceed as I tried to do with the Velikovsky business. One needs to collect the available evidence; that includes the opinions of experts, but with the caveat that one answers for oneself the question, to what degree does an expert have the competence to answer such a question in a categorical way? One looks for concordances and contradictions in the evidence; one weighs its validity by reference to original sources wherever possible. One asks oneself, where shall the onus of proof be put in this matter? Does the conventional wisdom, expressed in the experts' opinions, need to prove that the other side is wrong, or must the burden of proof lie elsewhere? In doing all this, one does well to remember that it is virtually impossible to establish a general negative, to show with certainty that any particular matter is utterly impossible.

This manner of proceeding involves effort. Often, not all the evidence is readily available. Original sources may be hard to come by. One finds oneself at times wading in a mass of technicalities of which one is quite ignorant, where one has ostensibly no competence to form an opinion about two opposing views. But it is worth the effort. When or if one arrives at a conclusion, he has established a belief that is in accord with his own set of other beliefs, and can feel comfortable with it. One becomes one's own expert, and it is more agreeable to have one's own informed opinion than to cling to the opinions of others merely because they are supposedly in a better position to form one.

And the quest itself offers a number of other possible, entirely unexpected rewards. My own interest in determining the truth about the Loch Ness monsters, for instance, led me to new insights about many other things, to new intellectual adventures, and—not least—made it possible for me to come to know the man to whom I have dedicated this book, Tim Dinsdale: a refreshing individual, well worth knowing for all sorts of reasons—for his courage and integrity, his enjoyment of life, his humor and his humility.

Then again, one may occasionally enjoy the human satisfaction of having been right, for the right reasons, when everyone else was wrong, perhaps even sarcastic. Or, one may find—as I did when I studied the Velikovsky affair—that much can become clear when one pays attention to things that were ignored by others. Giving much thought to one specific, apparently limited issue can lead to fruitful lines of thought, and even action, in quite other spheres.

As in all else, effort expended leads to rewards; activity is more satisfying than passivity; knowledge is better than ignorance. It is worth thinking for oneself.

References

[1] Abell, George O. *Science Books,* 12 (1976):126.

[2] ———. Review of *Scientists Confront Velikovsky. Physics Today,* 31 (no. 8) (Aug. 1978):56–59.

[3] ———, and Barry Singer, eds. *Science and the Paranormal.* New York: Charles Scribner's Sons, 1981.

[4] Albright, William F. "Retelling the Near East's Ancient History." *New York Herald Tribune* (20 Apr. 1952), sec. 6, p. 6.

[5] ———. "Velikovsky's Tour de Force of Legend, History and Psychoanalysis." *New York Herald Tribune Book Review* (29 May 1960), p. 4A.

[6] Aldiss, Brian W. *Billion Year Spree.* New York: Doubleday, 1973.

[7] Alexander, George. "Controversial Author, Scientists in Collision." *Los Angeles Times* (26 Feb. 1974), pt. II, p. 6.

[8] *American Behavioral Scientist* (Sept. 1963).

[9] ———. (Jan. 1964), p. 29. Unsigned editorial reply to Schenkman [363].

[10] ———. (Jan. 1964), p. 30. Unsigned editorial reply to de Camp [68].

[11] ———. (June 1964), p. 43. Unsigned editorial comment.

[12] *Analog Science Fiction/Science Fact.* "With Friends like These. . . ." (Jan. 1973), pp. 4–7, 177–78. Editorial.

[13] ———. "The Whole Truth." (Oct. 1974), pp. 5–11. Editorial.

[14] Anderson, John Lynde, and George W. Spangler. "Radiometric Dating: Is the 'Decay Constant' Constant?" *Pensée,* 9:31–33.

[15] Anderson, Poul. "Laputans and Lemurians." *Science,* 139 (1963):670, 672.

[16] Asimov, Isaac. "Worlds in Confusion." *Magazine of Fantasy and Science Fiction* (Oct. 1969). Reprinted in *The Stars in Their Courses.* New York: Doubleday, 1971, pp. 36–46.
[17] ———. "CP." *Analog Science Fiction/Science Fact* (Oct. 1974), pp. 38–50.
[18] Atwater, Gordon A. "Explosion in Science." *This Week* (2 Apr. 1950), pp. 10, 11, 20, 40.
[19] Aveni, Anthony F. Review of *Scientists Confront Velikovsky. Science,* 199 (Jan. 1978), pp. 288–89.
[20] Bailey, V. A. "Existence of Net Electric Charges on Stars." *Nature,* 186 (14 May 1960):508–10 and 189 (7 Jan. 1961):43–45.
[21] Barber, Bernard. "Resistance by Scientists to Scientific Discovery." *Science,* 134 (1961):596–602.
[22] Bargmann, V., and Lloyd Motz. "On the Recent Discoveries Concerning Jupiter and Venus." *Science,* 138 (1962):1350. Reprinted as Appendix I in de Grazia [78].
[23] Barzun, Jacques. *Science: The Glorious Entertainment.* New York: Harper and Row, 1964.
[24] Bass, Robert W. "Did Worlds Collide?" *Pensée,* 8:8–20.
[25] ———. "'Proofs' of the Stability of the Solar System." *Pensée,* 8:21–26.
[26] Beichman, Arnold. *Nine Lies about America.* New York: Library Press, 1972; New York: Pocket Books, 1973.
[27] Bell, Peter M. "Is the Velikovsky Era Ended?" *EOS,* 61 (no. 8) (19 Feb. 1980):89–90.
[28] Bernstein, Jeremy. *A Comprehensible World.* New York: Random House, 1967 .
[29] ———. "Scientific Cranks—How to Recognize One and What to Do Until the Doctor Arrives." *American Scholar,* 47 (1977–78):8–14.
[30] Bimson, John. "An Eighth-Century Date for Merenptah." *S.I.S. Review,* 3 (no. 2) (1978):57–59.
[31] Boffey, Philip M. "'Worlds in Collision' Runs into Phalanx of Critics." *Chronicle of Higher Education* (4 Mar. 1974), pp. 1, 7.
[32] Bok, Bart J., and Lawrence E. Jerome. *Objections to Astrology.* Buffalo, N.Y: Prometheus, 1975.
[33] Boring, Edwin G. "Psychological Factors in the Scientific Process." *American Scientist,* 42 (1954):639–45.
[34] ———. "The Validation of Scientific Belief." *Proceedings of the American Philosophical Society,* 96 (5 Oct. 1952):535–39.

[35] Breit, Harvey. "Talk with Mr. Velikovsky." *New York Times Book Review* (2 Apr. 1950), p. 12.
[36] Broad, William J. "Rewriting the History of the H-Bomb." *Science,* 218 (19 Nov. 1982):769–72.
[37] Brown, Harrison. "Venus and the Scriptures." *Saturday Review* (22 Apr. 1950), pp. 18–19.
[38] ———. *Scientific American* (Mar. 1956), pp. 127, 128, 130, 132.
[39] ———. *Scientific American* (May1956), pp. 12, 14, 16.
[40] Brown, Jack E. *Library Journal* (15 Nov. 1955), p. 2601.
[41] Bruce, C. E. R. *Pensée,* 3:53–54; 5:44–46.
[42] Burgstahler, Albert W. *Mankind Quarterly,* 7 (1966):121–23.
[43] ———. "The El-Amarna Letters and the Ancient Records of Assyria and Babylonia." *Pensée,* 5:13–15.
[44] ———. "The Nature of the Cytherean Atmosphere." *Pensée,* 6:24–30. Reprinted in *Velikovsky Reconsidered* [315].
[45] ———, and Euan W. Mackie. " 'Ages in Chaos' in the Light of C-14 Archaeometry." *Pensée,* 4:33–37, 50.
[46] Cantril, Hadley, *Harper's* (Oct. 1963), p. 14.
[47] Cazeau, Charles J., and Stuart D. Scott, Jr. *Exploring the Unknown.* New York: Plenum, 1979.
[48] Chedd, Graham, "Velikovsky in Chaos." *New Scientist* (7 Mar. 1974), pp. 624–25.
[49] *Choice* (July–Aug. 1976), p. 680. Unsigned book review.
[50] Christie, Ian R. *Times Literary Supplement* (6 Aug. 1976), p. 986.
[51] Cohen, Daniel. *Myths of the Space Age.* New York: Dodd, Mead, 1965.
[52] Cohen, I. Bernard. "Orthodoxy and Scientific Progress." *Proceedings of the American Philosophical Society,* 96 (5 Oct. 1952):505–12.
[53] ———. "An Interview with Einstein." *Scientific American* (July 1955), pp. 69–73.
[54] Coleman, W. "Abraham Gottlob Werner, vu par Alexander von Humboldt avec des notes de Georges Cuvier." *Sudhoffs Archiv für Geschichte der Medizin und der Naturwissenschaften,* 47 (1963):465–78. Reference drawn to my notice by Dr. Rachel Laudan.
[55] Conant, James B. *Science and Common Sense.* New Haven, Conn.: Yale University Press, 1951.
[56] Condon, Edward U. "Velikovsky's Catastrophe." *New Republic* (24 Apr. 1950), pp. 23–24.

[57] Converse, Philip E. *American Behavioral Scientist* (Nov. 1963) p. 22.
[58] Cook, Melvin A. *Pensée,* 3:55–57.
[59] Costello, Peter. *In Search of Lake Monsters.* London: Garnstone, 1974.
[60] Cowen, R. C. *Christian Science Monitor* (15 Apr. 1950), p. 18.
[61] Craig, Paul C. *Harper's* (Aug. 1951), p. 14.
[62] Crew, Eric W. *Pensée,* 5:47–48.
[63] Crouch, W. T. *American Behavioral Scientist* (Nov. 1963), p. 23.
[64] Davies, James C. *American Behavioral Scientist* (Jan. 1964), p. 30.
[65] Davis, Monte. Review of *Mankind in Amnesia. Discover* (Mar. 1982), pp. 88–89.
[66] Davis, Robert Gorham. "Velikovsky's World." *New Leader* (31 Jan. 1977), pp. 17–18.
[67] de Camp, L. Sprague. "Orthodoxy in Science." *Astounding Science-Fiction* (May 1954), pp. 116–29.
[68] ———. *American Behavioral Scientist* (Jan. 1964), p. 30.
[69] ———. *The Ragged Edge of Science.* Philadelphia: Owlswick, 1980.
[70] ———, and Catherine C. de Camp. *Spirits, Stars and Spells.* Canaveral, 1966.
[71] de Grazia, Alfred. *American Behavioral Scientist* (Sept. 1963), Foreword.
[72] ———. "Notes on 'Scientific' Reporting." *American Behavioral Scientist* (Oct. 1964), pp. 14–17.
[73] ———. "The Scientific Reception System." pp. 171–231 in reference 78.
[74] ———. "The Coming Cosmic Debate in the Sciences and Humanities: Revolutionary vs. Evolutionary Primevology." *Proceedings of Symposium,* Montreal, 10–12 Jan. 1975, pp. 21–40. Available from Saidye Bronfman Centre, 5170 Cote Saint Catherine's Road, Montreal, Quebec, Canada.
[75] ———. "Aphrodite—The Moon or Venus?" *S.I.S. Review,* 1 (no. 3) (1976):8–11.
[76] ———. "Paleo-Calcinology: Destruction by Fire in Prehistoric and Ancient Times." *Kronos,* 1 (no. 4) (1976): 25–36; 2 (no. 1):63–71.
[77] ———. *Chaos and Creation.* Princeton, London, Bombay: Metron Publications, 1981. The first in the Quantavolution Series; announced as forthcoming are *Moses and His Electric*

God; Homo Schizo: Human Nature and Its Origins; The Cosmic Heretic: I. Velikovsky; The Lately Tortured Earth; Solaria Binaria: Origins and History of the Solar System (with Earl R. Milton); *The Disastrous Love Affair of Moon and Mars; The Burning of Troy: Essays in Quantavolution; Divine Succession: The Life and Death of Gods.*

[78] ———, ed. *The Velikovsky Affair—The Warfare of Science and Scientism.* New York: University Books, 1966.

[79] Dean, Geoffrey. Letter, *Zetetic Scholar* (no. 6) (1980), p. 4.

[80] Dedmon, Emmett. "Offers Proof for Bible Miracles." *Chicago Sun-Times* (4 Apr. 1950).

[81] Deloria, Vine, Jr. "Myth and the Origin of Religion." *Pensée,* 9:45–50.

[82] Dempsey, David. "In and Out of Books: Venus Observed." *New York Times Book Review* (23 Apr. 1950), p. 8.

[83] ———. "In and Out of Books: $4.50 Question." *New York Times Book Review* (21 May 1950), p. 8.

[84] ———. "In and Out of Books: New Horse, Old Rider." *New York Times Book Review* (18 June 1950), p. 8.

[85] Dingle, Herbert. *Science at the Crossroads.* London: Martin Brian and O'Keefe, 1972.

[86] Dinsdale, Tim. *Loch Ness Monster.* London: Routledge and Kegan Paul, 1961, 1972, 1976, 1982.

[87] ———. *The Leviathans.* London: Routledge and Kegan Paul, 1966; 2d ed. entitled *Monster Hunt,* Washington, D.C.: Acropolis, 1972.

[88] ———. *The Story of the Loch Ness Monster.* London: Allan Wingate and Target-Universal-Tandem, 1973.

[89] ———. *Project Water Horse.* London and Boston: Routledge and Kegan Paul, 1975.

[90] Doermann, Humphrey. "Shapley Brands 'Worlds in Collision' a Hoax—Scientists' Attacks, Pressure Make Macmillan Call Off Publication." *Harvard Crimson* (25 Sept. 1950), pp. M1, 2.

[91] Donnelly, Ignatius. *Ragnarok: The Age of Fire and Gravel.* New York: D. Appleton, 1883.

[92] Doubleday and Company. Supplement of nine unnumbered pages bound with *Ages in Chaos* [410], beginning after p. 350; extracts from reviews of *Worlds in Collision* (the following citations are given in the sequence in which they appear):

A *Pageant*
B John J. O'Neill, *Book-of-the-Month Club News*
C *Everybody's Digest*
D *American Mercury*
E *New York Sunday News*
F George E. Sokolsky, *New York Journal and American*
G A. G. H. Deitz, *Boston Post*
H Lester Allen, *Boston Post*
I Kenneth Horan, *Dallas Times-Herald*
J Larry Howes, *Portland Journal*
K Lee Casey, *Rocky Mountain News*
L A. Branscombe, *Rocky Mountain News*
M Ward Moore, *National Jewish Post*
N *The Times,* London
O W. E. Garrison, *Christian Century*
P *Christian Herald*
Q Russell McCarthy, *Pasadena Star-News*

[93] ———. "A Report on the Velikovsky Controversy." *New York Times Book Review* (Aug. 1964), p. 17.
[94] Dudley, H. C. *Pensée,* 2:44.
[95] Edgerton, Franklin. *Harper's* (Aug. 1951), p. 14.
[96] Edmondson, Frank K. *Indianapolis Star* (9 Apr. 1950).
[97] Eichhorn-von Wurmb, H. K. *Pensée,* 8:47.
[98] Ellenberger, C. Leroy. "Comments on the Dialogue on Velikovsky." *Zetetic Scholar* no. 5 (1979), pp. 89–97 (see reference 475).
[99] ———. "Heretics, Dogmatists and Science's Reception of New Ideas." *Kronos,* 4 (no. 4) (1979):60–74; pt. 2, 5 (no. 4) (1980):48–69; pt. 3, 6 (no. 2) (1981):72–84; pt. 4, 6 (no. 4) (1981):71–84.
[99A] ———. Personal communications, 1983.
[99B] ———. "*Worlds in Collision* in Macmillan's Catalogues." *Kronos,* 9 (no. 2) (1984):46–57.
[100] ———, Lewis M. Greenberg, and Shane Mage. Letter, *Biblical Archaeology Review,* 6 (no. 3) (May/June 1980):10–12.
[101] Ellis, Albert. *Humanistic Psychotherapy.* New York: McGraw-Hill, 1973.
[102] ———, and Robert A. Harper. *A New Guide to Rational Living.* North Hollywood, Calif.: Wilshire, 1975.

[103] Engel, Leonard. "Comets and Cataclysms." *New York Times Book Review* (20 Nov. 1955), p. 28.
[104] Evans, Christopher. *Cults of Unreason.* New York: Delta-Dell Publishing, 1975.
[105] Fabun, Don. "Another Controversial Book by Velikovsky." *San Francisco Chronicle* (9 June 1952), pp. 18, 21.
[106] Fair, Charles. *The New Nonsense.* New York: Simon and Schuster, 1974.
[107] Ferté, Thomas. "A Record of Success." *Pensée,* 1:11–15, 23.
[108] Fessler, Aaron F. *Library Journal* (15 May 1966), p. 2507.
[109] Forrest, Bob. *Velikovsky's Sources.* Privately published by Bob Forrest, 53 Bannerman Avenue, Prestwich, Manchester M25 5DR, England; parts 1 and 2 (1981); 3, 4, and 5 (1982); 6 and volume of index and notes (1983).
[110] Gammon, Geoffrey. "Newton's History." *S.I.S. Review,* 1 (no. 4) (1977):16.
[111] Gamow, George. *Mr. Tompkins in Wonderland, or Stories of c, G, and h.* New York: Macmillan, 1945.
[112] Gardner, Martin. "The Hermit Scientist." *Antioch Review* 10 (1950):447–57.
[113] ———. *In the Name of Science.* New York: G. P. Putnam's Sons, 1952. References are to the 2d ed., *Fads and Fallacies in the Name of Science.* New York: Dover, 1957.
[114] ———. *Science—Good, Bad and Bogus.* Buffalo, N.Y.: Prometheus, 1981.
[115] Garrison, W. E. "Chaos in the Cosmos." *Christian Century* (7 June 1950), pp. 703–4.
[116] Gell-Mann, Murray. "How Scientists Can Really Help." *Physics Today* (May 1971), pp. 23–25.
[117] Gillette, Robert. "Velikovsky: AAAS Forum for a Mild Collision." *Science,* 183 (1974):1059–62. Reprinted in *Current* (Plainfield, Vt.) (May 1974), pp. 38–44.
[118] Goldsmith, Donald. *Scientists Confront Velikovsky.* Introduction, pp. 19–28. See [119].
[119] Goldsmith, Donald, ed. *Scientists Confront Velikovsky.* Ithaca, N.Y.: Cornell University Press, 1977.
[120] Good, I. J. University Distinguished Professor, Virginia Polytechnic Institute and State University; private communication, 1979.
[121] ———. "A Subjective Evaluation of Bode's Law and an 'Ob-

jective' Test for Approximate Numerical Rationality." *American Statistical Association Journal,* 64 (Mar. 1969): 23–66, especially pp. 28–29.

[122] Goodavage, Joseph F. "An Interview with Carl Sagan." *Analog Science Fiction/Science Fact* (Aug. 1976), pp. 92–101.

[123] Goode, John. "Keeping an Open Mind on Velikovsky." *Sydney Morning Herald* (Australia) (26 Oct. 1976), p. 7.

[124[Goodman, George, Jr. *New York Times* (19 Nov. 1979), p. D9.

[125] Gordon, Theodore J. *Ideas in Conflict.* New York: St. Martin's, 1966.

[126] Goudsmit, Samuel A. "It Might as Well Be Spin." *Physics Today* (June 1976), pp. 40–43.

[127] Gould, Stephen Jay. "Velikovsky in Collision." *Natural History,* 84 (Mar. 1975):20, 24, 26.

[128] ———. "Reverend Burnet's Dirty Little Planet." *Natural History,* 84 (Apr. 1975):26–28.

[129] ———. "Evolution as Fact and Theory." *Discover* (May 1981), pp. 34–37.

[130] Greenberg, Lewis M. *Pensée,* 3:36–37.

[131] ———. "Atlantis." *Pensée,* 6:51–54.

[132] ———. "Sagan's Folly." Pt. I, *Kronos,* 3 (no. 2) (1977):62–82.

[133] ———. Editorial statement. *Kronos,* 3 (no. 4) (1978):2.

[134] ———. Editor's note. *Kronos,* 5 (no. 1) (1979):3.

[135] Griffard, David. "Myth, Mandala and the Collective Unconscious." *Kronos,* 1 (no. 1) (1975):27–32.

[136] Gruenberger, Fred J. "A Measure for Crackpots." *Science,* 145 (25 Sept. 1964):1413–15.

[137] Guthrie, Warren. "Books, Civilization, and Science." *Science,* 113 (1951):429–31.

[138] Hadas, Moses. "What We Remember." *Reporter* (9 Apr. 1964), pp. 45–46.

[139] Haldane, J. B. S. "St. Quetzalcoatl and St. Fenris." *New Statesman and Nation* (11 Nov. 1950), pp. 432–33.

[140] ———. Reply to letter by Velikovsky. *New Statesman and Nation* (3 Feb. 1951).

[141] *Harper's.* "Personal and Otherwise." (Jan. 1950), pp. 6, 8.

[142] ———. "Sages in Chaos." (June 1951), pp. 9–11.

[143] ———. Editorial note. (June 1951), p. 51.

[144] Hawkins, Gerald S. *Stonehenge Decoded.* London: Fontana/Collins, 1970.

[145] ———. *Beyond Stonehenge.* New York: Harper and Row, 1973.
[146] Herget, Paul, "It Is Just as Barnum Said." *Cincinnati Enquirer* (1 Apr. 1950), p. 7.
[147] Hering, Daniel W. *Foibles and Fallacies of Science.* New York: Van Nostrand, 1924.
[148] Hitching, Francis. *The Mysterious World—An Atlas of the Unexplained.* New York: Holt, Rinehart and Winston, 1979.
[149] Holbrook, John, Jr. "Outline of the 1st Millennium B.C. Following Immanuel Velikovsky's Reconstruction of Ancient History." *Pensée,* 4:centerfold (a chart).
[150] Hornsby, Henry. *Louisville Courier-Journal* (26 June 1977).
[151] Hulburt, E. O. Chapter X, "The Upper Atmosphere." Pp. 492–572, in *Terrestrial Magnetism and Electricity* (Physics of the Earth, vol. 8), ed. John Adams Fleming. New York: McGraw-Hill, 1939.
[152] Humphreys, W. J. *Physics of the Air.* 2d ed. New York: McGraw-Hill, 1929.
[153] Hunt, Ben. *Catholic World,* 171:477.
[154] *Industrial Research.* Unsigned news item (Dec. 1973), pp. 25–26.
[155] *Irish Astronomical Journal.* "Queries and Answers." 1 (1951):167–68.
[156] Isaacson, Israel M. "Carbon-14 Dates and Velikovsky's Revision of Ancient History." *Pensée,* 4:26–32.
[157] ———. "Applying the Revised Chronology." *Pensée,* 9:5–20, 33.
[158] James, Peter. "Aphrodite—The Moon or Venus?" *S.I.S. Review,* 1 (no. 1) (1976):2–7.
[159] ———. "A Critique of 'Ramses II and His Time.'" *S.I.S. Review,* 3 (no. 2) (1978):48–55.
[160] Jastrow, Robert. "Hero or Heretic?" *Science Digest Special* (Sept./Oct. 1980), pp. 92–96.
[161] Jeans, Sir James. *An Introduction to the Kinetic Theory of Gases.* New York and Cambridge: Macmillan and Cambridge University Press, preface dated 1940.
[162] Jerome, Lawrence E. "Astrology: Magic or Science?" Pp. 43–62 in *Objections to Astrology.* Buffalo, N.Y.: Prometheus Books, 1975. Reprinted from *The Humanist,* 35 (no. 5) (Sept./Oct. 1975).

[163] Jones, Sir Harold Spencer. "False Trail." *Spectator* (22 Sept. 1950), p. 321.
[164] ———. "The Flying Saucer Myth." *Spectator* (15 Dec. 1950), pp. 686–87.
[165] Jordan, David Starr. *The Higher Foolishness.* Indianapolis: Bobbs-Merrill, 1927.
[166] Jueneman, Frederic B. "Velikovsky." *Industrial Research* (Mar. 1973), pp. 40–44.
[167] ———. "The Search for Truth." *Analog Science Fiction/Science Fact* (Oct. 1974), pp. 25–37.
[168] ———. "pc (Psycho-Ceramics)." *Kronos,* 1 (no. 3) (1975):73–83.
[169] ———. "A Kick in the AAAS." *Industrial Research* (Aug. 1976), p. 9.
{170] ———. "The Old Warrior Does It Again." *Industrial Research* (Feb. 1977), p. 13.
[171] ———. "Velikovsky Gone, Questions Remain." *Industrial Research/Development* (Jan. 1980), pp. 54, 56, 58, 60.
[172] Juergens, Ralph E. "Minds in Chaos." Pp. 7–49 in de Grazia [78].
[173] ———. "Aftermath to Exposure." Pp. 50–79 in de Grazia [78].
[174] ———. "A Rejoinder to Motz." *Yale Scientific Magazine,* 41 (Apr. 1967):16, 30.
[175] ———. "Reconciling Celestial Mechanics and Velikovskian Catastrophism." *Pensée,* 2:6–12. Reprinted in *Velikovsky Reconsidered* [315].
[176] ———. *Pensée,* 3:52–53, 54–55, 57–58; 5:43–44, 48–50.
[177] ———. "Electrical Discharges and the Transmutation of Elements." *Pensée,* 8:45–46.
[178] ———. "Of the Moon and Mars." Pt. 1, *Pensée,* 9:21–30.
[179] Kaempffert, Waldemar. "The Tale of Velikovsky's Comet." *New York Times Book Review* (2 Apr. 1950), pp. 1, 16.
[180] ———. "Solomon, the Queen of Sheba, and the Egypt of the Exodus." *New York Times Book Review* (20 Apr. 1952), p. 23.
[181] ———. *New York Times Book Review* (7 May 1950), p. 17.
[182] Kallen, Horace M. *Yale Scientific Magazine,* 41 (1967):30.
[183] ———. "Shapley, Velikovsky and the Scientific Spirit." *Pensée,* 1:36–40. Reprinted in *Velikovsky Reconsidered* [315].
[184] ———. *Creativity, Imagination, Logic.* New York: Gordon and Breach, 1973.

[185] Kazin, Alfred. "On the Brink." *New Yorker* (29 Apr. 1950), pp. 103–5.
[186] Keister, J. C., and Andrew Hamilton. "An Alternative to the Ejection of Venus from Jupiter." *S.I.S. Review,* 3 (no. 2) (1978):45–48.
[187] Kennedy, John L. "An Evaluation of Extra-Sensory Perception." *Proceedings of the American Philosophical Society,* 96 (1952):513–18.
[188] Kervran, Louis C. *Biological Transmutations.* Trans. Michel Abehsera. Binghamton, N.Y.: Swan House Publishing, 1972.
[189] *Kirkus Reviews.* (15 Dec. 1949), p. 689. Unsigned book review.
[190] ———. (1 Feb. 1976), p. 177. Unsigned book review.
[191] Knight, Damon. *Charles Fort—Prophet of the Unexplained.* New York: Doubleday, 1970.
[192] Kogan, S. F. Letter re Sagan versus Velikovsky. *Physics Today* (Sept. 1980), pp. 97–98.
[193] Kogan, Shulamit. "Sagan vs. Sagan." *Kronos,* 6 (no. 3) (1981): 34–41.
[194] Kohlenberg, Arthur. *Time* (3 Apr. 1950), p. 4.
[195] Kolodiy, George. "Velikovsky: Paradigms in Collision." *Bulletin of the Atomic Scientists* (Feb. 1975), pp. 36–38.
[196] Kornberg, Arthur. Quoted in *Discover* (Apr. 1981), p. 62.
[197] Krogman, Wilton H. "Author Finds a Time-Dislocation and Takes Ancient History for a Ride." *Chicago Sunday Tribune* (20 Apr. 1952), pt. 4, p. 4.
[198] ———. "More Velikovsky Revisions." *Chicago Sunday Tribune (Magazine of Books)* (27 Nov. 1955), p. 2.
[199] *Kronos—A Journal of Interdisciplinary Synthesis.* P. O. Box 343, Wynnewood, Pa. 19096; 4 issues per year, beginning with vol. 1 (no. 1) (Spring 1975).
[200] ———. Editor's preface to Jueneman [168], p. 73.
[201] ———. *Velikovsky and Establishment Science.* Special issue, vol. 3 (no. 2) (1977); also available in hardcover.
[202] ———. 3 (no. 2) (1977):139.
[203] Kruskal, Martin. *Pensée,* 3:51–52; 5:42–43.
[204] Kuhn, Thomas S. *The Structure of Scientific Revolutions.* Chicago: University of Chicago, 1962. 2d ed., 1970.
[205] Kuong, Wong Kee. "The Synthesis of Manna." *Pensée,* 3:45–46.
[206] Kurtz, Paul, and Lee Nisbet. "Are Astronomers and Astrophysicists Qualified to Criticize Astrology?" *Zetetic,* 1 (no. 1) (1976):47–52.

[207] Lafleur, Laurence J. "Cranks and Scientists." *Scientific Monthly* (Nov. 1951), pp. 284–90.
[208] Langmuir, I. "Pathological Science." Report no. 68-C-035, General Electric Research and Development Center, Schenectady, N.Y. (Apr. 1968), 13 pp.
[209] Larrabee, Eric. "The Day the Sun Stood Still." *Harper's* (Jan. 1950), pp. 19–26.
[210] ———. *Reporter* (11 Apr. 1950).
[211] ———. "Scientists in Collision: Was Velikovsky Right?" *Harper's* (Aug. 1963), pp. 48–55.
[212] ———. "A Comment on Dr. Menzel's Rejoinder." *Harper's* (Aug. 1963), p. 87.
[213] Latham, Harold S. *My Life in Publishing.* New York: E. P. Dutton, 1965.
[214] Latourette, K. S., et al. *American Journal of Science,* 248 (1950):584–89.
[215] Layzer, David. *Harper's* (Mar. 1950), p. 19.
[216] Leakey, Richard E. Quoted in Cheryl M. Fields, *Chronicle of Higher Education* (14 Apr. 1982), p. 5.
[217] Lear, John. "The Heavens Burst." *Collier's* (25 Feb. 1950), pp. 24, 42–45.
[218] ———. "World on Fire." *Collier's* (25 Mar. 1950), pp. 24, 82–85.
[219] Ley, Willy. *Book Week* (11 July 1965), p. 14.
[220] Libby, W. F. "The Radiocarbon Dating Method." *Pensée,* 4:7–11.
[221] Longwell, Chester R. *Science,* 113 (1951):418.
[222] Lowery, Malcolm. "What's in a Name?—Venus 'The Newcomer.'" *S.I.S. Review,* 5 (no. 2) (1980/81):46–49.
[223] Lowery, R. M. *Times Literary Supplement* (27 Aug. 1976), p. 1054.
[224] Ludwig, George R. *Time* (3 Apr. 1950), p. 4.
[225] Lundberg, George A. *American Behavioral Scientist* (Nov. 1963), p. 22.
[226] Lybeck, A. H. *Harper's* (Mar. 1950), p. 18.
[227] Maccoby, Hyam. *Times Literary Supplement* (9 July 1976), p. 852.
[228] ———. "Velikovsky's Egypt." *Listener* (24 Feb. 1977), pp. 252–53.
[229] Mackal, Roy P. *The Monsters of Loch Ness.* Chicago: Swallow, 1976.

[230] MacKie, Euan. "A Challenge to the Integrity of Science?" *New Scientist* (11 Jan. 1973), pp. 76–77.
[231] ———. "A Quantitative Test for Catastrophic Theories." *Pensée*, 3:6–9. See also Burgstahler and MacKie [45].
[232] ———. "Megalithic Astronomy and Catastrophism." *Pensée*, 10:5–20.
[233] MacNamara, Charles H. "The Persecution and Character Assassination of Immanuel Velikovsky as Performed by the Inmates of the Scientific Establishment." *Philadelphia Magazine* (Apr. 1968), pp. 63–65, 92–98, 103.
[234] Mann, Alfred E. "Dr. Velikovsky and Edgar Cayce." *Searchlight*, 17 (no. 6):1–8; published by Association for Research and Enlightenment, Inc., Virginia Beach, Va.
[235] Margerison, Tom. *New Scientist* (22 July 1976), p. 187.
[236] Margolis, Howard. "Velikovsky Rides Again." *Bulletin of the Atomic Scientists* (Apr. 1964), pp. 38–40.
[237] Marks, David, and Richard Kammann. *The Psychology of the Psychic.* Buffalo, N.Y.: Prometheus, 1980.
[238] Mather, Kirtley F. *American Scientist*, 38 (1950):474, 476.
[239] Maultsby, Maxie C. *Help Yourself to Happiness through Rational Self-Counseling.* Boston: Marlborough House, Herman Publishing, 1975; third printing published by Institute for Rational Living, Inc., New York.
[240] May, Joseph. "A Call to Action." *Pensée*, 2:47–48.
[241] ———. Response to comments [475]. *Zetetic Scholar*, no. 5 (1979), pp. 78–84.
[242] ———. Reply to Dean [79]. *Zetetic Scholar*, no. 7 (Dec. 1980), p. 121.
[243] McCrea, W. H. *Proceedings of the Royal Society*, A256 (1960):245.
[244] McCurdy, Patrick P. Editorial. *Chemical and Engineering News* (9 Apr. 1973), p. 1.
[245] McTighe, Thomas P. *Best Sellers* (15 Aug. 1950), pp. 83–84.
[246] Menzel, Donald H. *Harper's* (Mar. 1950), p. 18.
[247] ———. "The Debate over Velikovsky." *Harper's* (Dec. 1963), pp. 83–86.
[248] Michelson, Irving. "Mechanics Bears Witness." *Pensée*, 7:15–21.
[249] ———. *Science*, 185 (1974):207–8.
[250] ———. "Velikovsky's Catastrophism." *Analog Science Fiction/Science Fact* (June 1975), pp. 65–76.

[251] Miller, Julius Sumner. *Harper's* (Aug. 1951), p. 14.
[252] Moore, Brian. "Sages in Chaos—The British Side of the Velikovsky Affair." *S.I.S. Review,* 1 (no. 2) (1976):18–19.
[253] ———. *S.I.S. Review,* 1 (no. 3) (1976):25.
[254] ———. "Bookshelf—Defenders of the Faith." *S.I.S. Review,* 2 (no. 4) (1978):99.
[255] Moore, J. B. *Times Literary Supplement* (9 July 1976), p. 852.
[256] Moore, Patrick. *Can You Speak Venusian?* New York: W. W. Norton, 1972.
[257] Morrison, David. *Physics Today* (Feb. 1972), pp. 72–73.
[258] ———. Replies to May [241], Ellenberger [98], and Ransom [474]. *Zetetic Scholar,* no. 7 (Dec. 1980), pp. 128–30.
[259] ———. Letter. *Physics Today* (Apr. 1981), pp. 72–73.
[260] Motz, Lloyd. *Harper's* (Oct. 1963), pp. 12, 14.
[261] ———. "Velikovsky—A Rebuttal." *Yale Scientific Magazine,* 41 (Apr. 1967):12–13.
[262] ———. *The Universe—Its Beginning and End.* New York: Charles Scribner's Sons, 1975.
[263] Mowles, Thomas. "Radiocarbon Dating and Velikovskian Catastrophism." *Pensée,* 4:19–25.
[264] Mullen, William. "The Center Holds." *Pensée,* 1:32–34. Reprinted in *Velikovsky Reconsidered* [315].
[265] ———. "A Reading of the Pyramid Texts." *Pensée,* 3:10–16.
[266] ———. "The Mesoamerican Record." *Pensée,* 9:34–44.
[267] Murdock, George P. *Harper's* (Mar. 1950), p. 18.
[268] *Nature.* "Science beyond the Fringe." 248 (1974):541.
[269] Neugebauer, O. *Isis,* 41 (1950):245–46.
[270] Newman, Edwin. *Strictly Speaking.* New York: Bobbs-Merrill, 1974.
[271] *Newsweek.* "The Universe: This Colliding World." (9 Jan. 1950), pp. 16, 19.
[272] ———. "Academic Freedom: Professors as Suppressors." (3 July 1950), pp. 15, 16, 19.
[273] ———. "Scientists in Collision." (25 Feb. 1974), pp. 58, 61.
[274] Nichols, Herbert B. "The Velikovsky Excursion." *Christian Science Monitor* (29 Mar. 1950), p. 18.
[275] Nieto, M. M. *The Titius-Bode Law of Planetary Distances: Its History and Theory.* Philadelphia: Pergamon, 1972. Quoted from Nieto [276:7n].
[276] ———. *Pensée,* 8:5–7.

[277] North, John. "Venus, by Jupiter!" *Times Literary Supplement* (25 June 1976), pp. 770–71.
[278] O'Neill, John J. "Atomic Energy Charging Globe Held Able to Erupt at Any Time." *New York Herald Tribune* (11 Aug. 1946), sec. II–IV, p. 10.
[279] Öpik, E. "'Worlds in Collision' and the Peril of Credulity." *Irish Astronomical Journal,* 10 (1972): 293–95.
[280] Orbell, John M. *American Political Science Review,* 61 (1967): 180–81.
[281] Orlinsky, Harry M. "Chaos in Ages." *Jewish Bookland* (Sept. 1952).
[282] Orwell, George. *Nineteen Eighty-Four.* London: Secker and Warburg, 1949.
[283] Osborn, Andrew D. *Library Journal,* 77 (1952):1076–77.
[284] Oursler, Fulton. "Why the Sun Stood Still." *Reader's Digest* (Mar. 1950), pp. 139–48.
[285] Parry, Thomas Alan. "The New Science of Immanuel Velikovsky." *Kronos,* 1 (no. 1) (1975):3–20.
[286] Paterson, A. M. "Giordano Bruno's View on the Earth without a Moon." *Pensée,* 3:46–47. Reprinted in *Velikovsky Reconsidered* [315].
[287] Payne-Gaposchkin, Cecilia. "Nonsense, Dr. Velikovsky." *Reporter* (14 Mar. 1950), pp. 37–40.
[288] ———. *Reporter* (11 Apr. 1950).
[289] ———. *Popular Astronomy* (June 1950), pp. 278–86.
[290] ———. "Worlds in Collision." *Proceedings of the American Philosophical Society,* 96 (1952):519–25.
[291] *Pensée* (defunct periodical). Between 1972 and 1975 *Pensée* published ten issues (numbered 1 through 10) in a series entitled "Immanuel Velikovsky Reconsidered." At the date of this writing, those issues were available from the Society for Interdisciplinary Studies [346]. A number of the articles were reprinted in book form [315].
[292] ———. 1:5. Unsigned biographical note about Velikovsky.
[293] ———. 2:16–17, 48. Unsigned editorial reply to Straka [391].
[294] ———. "Velikovsky on the Formation of Coal." 2:19–21. Unsigned.
[295] ———. 2:30–32. Notes by Editor and Cyrus Gordon, and quotation from C. D. Darlington.

[296] ———. 2:37. News item.
[297] ———. 3:4. Unsigned editorial.
[298] ———. 3:36. News item.
[299] ———. 3:39. News item.
[300] ———. 5:22–24. News item.
[301] ———. "Ash." 6:5–19. Collection of letters with editorial comments.
[302] ———. 7:23.
[303] ———. 7:28. Editorial report on arrangement of AAAS symposium.
[304] ———. 7:34. Editorial report on Velikovsky at AAAS symposium.
[305] ———. 7:37. Editorial report on Sagan at AAAS symposium.
[306] ———. 7:39. Comment by Juergens on Sagan at AAAS symposium.
[307] ———. 7:47. News item.
[308] ———. 7:48. News item.
[309] ———. 8:7. Report from symposium at McMaster University.
[310] ———. 8:42.
[311] ———. 10:40–41.
[312] ———. 10:43. Editorial comment.
[313] ———. 10:45. Report from meeting of Philosophy of Science Association.
[314] ———. News items from issues 1 to 10.
[315] *Pensée. Velikovsky Reconsidered.* New York: Doubleday, 1976.
[316] Perrow, Charles. *American Behavioral Scientist* (Jan. 1964), p. 29.
[317] Pfeiffer, John. "Illiteracy Triumphant." *Science,* 114 (13 July 1951):47.
[318] Plant, Albert F. Editorial. *Industrial Research* (May 1974), p. 9.
[319] Plummer, William T. "Venus Clouds: Test for Hydrocarbons." *Pensée,* 6:20–21. Reprinted from *Science,* 163 (1969):1191–92; reprinted in *Velikovsky Reconsidered* [315].
[320] Polanyi, Michael. "The Republic of Science." *Minerva,* 1 (1962):54–73.
[321] ———. "The Growth of Science in Society." *Minerva,* 5 (1967):533–45.
[322] Pope, Alexander. *An Essay on Criticism.* Lines 215–18.
[323] *Popular Science Monthly* (July 1876), p. 290.
[324] *Publishers' Weekly,* 157 (25 Feb. 1950):1120.

[325] ———. 157 (17 June 1950):2645.
[326] ———. 157 (24 June 1950):2739.
[327] ———. 209 (26 Jan. 1976):286.
[328] ———. "Mankind in Amnesia." (6 Nov. 1981), p. 74.
[329] Randi, James. *Flim-Flam.* Philadelphia and New York: Lippincott and Crowell, 1980.
[330] Ransom, C. J. Executive Director, Cosmos and Chronos, P.O. Box 12807, Fort Worth, Tex. 76116; Ransom holds a Ph.D. in physics and has worked in the Convair Aerospace Division of General Dynamics.
[331] ———. *The Age of Velikovsky.* Glassboro, N.J.: Kronos Press, 1976.
[332] ———. "Sagan's Appendices: A Quick Appendectomy." *Kronos,* 3 (no. 2) (1977):135–39.
[333] ———. On David Morrison's commentary [475]. *Zetetic Scholar,* no. 6 (1980), pp. 141–42.
[334]———, and L. H. Hoffee. "The Orbits of Venus." *Pensée,* 3:22–25.
[335] Reade, M. G. "Manna as a Confection." *S.I.S. Review,* 1 (no. 2) (1976):9–13, 25.
[336] Reber, Grote. *Science,* 140 (28 June 1963), p. 1362.
[337] Research Communications Network is now defunct.
[338] Riddick, Thomas M. "Dowsing. . . ." *Proceedings of the American Philosophical Society,* 96 (1952):526–34.
[339] Rines, Robert H., et al. "Search for the Loch Ness Monster." *Technology Review* (Mar.–Apr. 1976), pp. 25–40.
[340] Robertson, Miranda. "Velikovsky in the Open." *Nature,* 248 (15 Mar. 1974):190.
[341] Rose, Lynn E. "The Censorship of Velikovsky's Interdisciplinary Synthesis." *Pensée,* 1:29–31. Reprinted in *Velikovsky Reconsidered* [315].
[342] ———. "Could Mars Have Been an Inner Planet?" *Pensée,* 1:42–43. Reprinted in *Velikovsky Reconsidered* [315].
[343] ———. "Babylonian Observations of Venus." *Pensée,* 3:18–22. Reprinted in *Velikovsky Reconsidered* [315].
[344] ———. "Some Good Advice." *Kronos,* 3 (no. 2) (1977):iii–v.
[345] ———, and Raymond C. Vaughan. "Velikovsky and the Sequence of Planetary Orbits." *Pensée,* 8:27–34. Reprinted in *Velikovsky Reconsidered* [315].
[346] *S.I.S. Review.* Beginning with vol. 1, no. 1, Jan. 1976; published by Society for Interdisciplinary Studies, ed. Ber-

nard T. Prescott, 12 Dorsett Road, London SW19 3HA, England.
[347] ———. 1 (no. 2):21. Comment by Ragnar Forshufvud.
[348] ———. 1 (no. 3):1.
[349] ———. 1 (no. 3):16. Report of talk by Alfred de Grazia.
[350] ———. 1 (no. 3):18. News item.
[351] ———. 1 (no. 4):3. Report of talk by Alfred de Grazia.
[352] ———. 1 (no. 4):4. Reported from discussion at society meeting.
[353] Sagan, Carl. "An Analysis of 'Worlds in Collision.'" Paper given at meeting of American Association for the Advancement of Science, San Francisco, 25 Feb. 1974. Extracts were reported in *Pensée,* 7:36–38, 41–42; a 1976 version was circulated widely [132].
[354] ———. "If There Are Any, Could There Be Many?" *Nature,* 264 (9 Dec. 1976):497.
[355] ———. Letter. *Humanist,* 36 (no. 1) (Jan./Feb. 1976):2.
[356] ———. Chapter 2, "An Analysis of *Worlds in Collision.*" Pp. 41–104, in *Scientists Confront Velikovsky* [119].
[357] ———. "Analysis of 'Worlds in Collision.'" *Humanist* (Nov./Dec. 1977), pp. 11–21.
[358] ———. "Venus and Dr. Velikovsky." Pp. 81–127, 317–27, in *Broca's Brain.* New York: Random House, 1979.
[359] ———. "A Scientist Looks at Velikovsky's 'Worlds in Collision.'" *Biblical Archaeology Review,* 6 (no. 1) (Jan./Feb. 1980): 40–51.
[360] ———. *Cosmos.* New York: Random House, 1980.
[361] ———. Chapter 15, "An Analysis of *Worlds in Collision.*" Pp. 223–52, in Abell and Singer [3].
[362] *Saturday Evening Post.* "The 1950 Silly Season Looks Unusually Silly." (18 Nov. 1950), pp. 10–12.
[363] Schenkman, Alfred S. *American Behavioral Scientist* (Jan. 1964), p. 29.
[364] *Science.* "The Science Court Experiment: An Interim Report." 193 (1976):653–56.
[365] *Science Newsletter.* "Theories Denounced." 57 (1950):119.
[366] Searle, John R. *The Campus War.* New York: Penguin, 1972.
[367] Shaw, George Bernard. *Everybody's Political What's What?* New York: Dodd, Mead, 1944.
[368] Sherrerd, Chris S. "Venus' Critical Orbit." *Pensée,* 1:43.
[369] ———. "Gyroscopic Precession and Celestial Axis Displace-

ment." *Pensée,* 5:31–33. Reprinted in *Velikovsky Reconsidered* [315].
[370] Shurkin, Joel N. "Blaze of Glory for an Old Man." *Philadelphia Inquirer* (27 Feb. 1974), pp. 1–2A.
[371] Sieff, Martin. "The Two Jehorams." *S.I.S. Review,* 2 (no. 3) (1977/78):86–90.
[372] *Skeptical Inquirer.* Published by the Committee for the Scientific Investigation of Claims of the Paranormal, Box 229, Central Park Station, Buffalo, N.Y. 14215.
[373] Sladek, John. *The New Apocrypha.* London: Hart-Davis, MacGibbon, 1973.
[374] Smith, Peter J. "Velikovsky." *New Society* (1 July 1976), p. 35.
[375] Smith, P. V. *Science,* 116 (1952):437.
[376] *Society for Scientific Exploration.* See *Science,* 216 (23 Apr. 1982):360; *Smithsonian,* 13 (no. 5) (Aug. 1982):26, 28, 30.
[377] Sorensen, Herbert C. "The Ages of Bristlecone Pine." *Pensée,* 4:15–18.
[378] Stecchini, Livio C. "The Inconstant Heavens." Pp. 80–126 in de Grazia [78].
[379] ———. "Astronomical Theory and Historical Data." Pp. 127–70 in de Grazia [78].
[380] Stent, Gunther S. "Prematurity and Uniqueness in Scientific Discovery." *Scientific American* (Dec. 1972), pp. 84–93.
[381] Stephanos, Robert. "Ben Franklin's Town: Where 'Lightning' Strikes Twice." *Yale Scientific Magazine,* 41 (Apr. 1967): 26–28.
[382] Stetson, R. H. *Harper's* (Mar. 1950), p. 18.
[383] Stevens, Joseph H. *Harper's* (Mar. 1950), p. 18.
[384] Stewart, George R. "A Tale of Erratic Planets and Lost Comets." *San Francisco Chronicle, This World* (2 Apr. 1950), p. 24.
[385] Stewart, John Q. "Disciplines in Collision." *Harper's* (June 1951), pp. 57–63.
[386] Stiebing, William H., Jr. "A Criticism of the Revised Chronology." *Pensée,* 5:10–12.
[387] ———. "Rejoinder to Velikovsky." *Pensée,* 10:24–26.
[388] Story, Ronald. *The Space-Gods Revealed.* New York: Harper and Row, 1976.
[389] Stove, David. "Velikovsky in Collision." *Quadrant* (Oct.–Nov. 1964), pp. 35–44.
[390] ———. "The Scientific Mafia." *Pensée,* 1:6–8, 49. Reprinted in *Velikovsky Reconsidered* [315].

[391] Straka, W. C. "Velikovsky: Science or Anti-Science?" *Pensée,* 2:13, 15.

[392] Strausz-Hupé, Robert. *American Behavioral Scientist* (Nov. 1963), p. 22.

[393] Struve, Otto. "Copernicus? Who Was He?" *New York Herald Tribune Book Review* (2 Apr. 1950), p. 4.

[394] Suleiman, Jo-Ann D. *Library Journal,* 102 (1977):199.

[395] Sullivan, Walter. "The Velikovsky Affair." *New York Times* (2 Oct. 1966), sec. 4, p. E7.

[396] ———. *Continents in Motion.* New York: McGraw-Hill, 1974.

[397] Sutherland, Carter. "China's Dragon." *Pensée,* 6:47–50.

[398] Talbott, Stephen L. "Velikovsky at Harvard." *Pensée,* 1:47–49.

[399] Thomsen, Dietrick E. "Velikovsky Lives Again." *New York Times Book Review* (17 Apr. 1977), pp. 10–11.

[400] *Time.* "Venus on the Loose." (13 Mar. 1950), pp. 72–74.

[401] *Times Literary Supplement.* "Scientific Speculation." (22 Sept. 1950), p. 591.

[402] ———. "Myth and Moonshine." (20 Jan. 1961), p. 43.

[403] Treash, Robert. "Magnetic Remanence in Lunar Rocks." *Pensée,* 1:21–23. Reprinted in *Velikovsky Reconsidered* [315].

[404] ———. "Is Asteroid Toro a Remnant of Comet-Planet Collision?" *Pensée,* 2:43–44.

[405] Uhlan, Edward. *The Rogue of Publishers' Row.* New York: Exposition, 1956.

[406] Velikovsky, Immanuel. *Theses for the Reconstruction of Ancient History from the End of the Middle Kingdom to the Advent of Alexander the Great.* No. 3 in the series *Scripta Academica Hierosolymitana.* New York and Jerusalem, 1945.

[407] ———. *Cosmos without Gravitation: Attraction, Repulsion, and Circumduction in the Solar System.* No. 4 in the series *Scripta Academica Hierosolymitana.* New York and Jerusalem, 1946.

[408] ———. *Worlds in Collision.* Page references are to the Doubleday hardcover ed. New York, 1950.

[409] ———. *Worlds in Collision.* Paperback ed. New York: Pocket Books, 1977.

[410] ———. *Ages in Chaos.* Vol. 1: *From the Exodus to King Akhnaton.* New York: Doubleday, 1952.

[411] ———. *Earth in Upheaval.* New York: Doubleday, 1955. Page references are to the Dell Laurel ed. New York, 1968.

[412] ———. *Oedipus and Akhnaton.* New York: Doubleday, 1960.

[413] ———. *Peoples of the Sea.* New York: Doubleday, 1977.
[414] ———. *Ramses II and His Time.* New York: Doubleday, 1978.
[415] ———. *Mankind in Amnesia.* New York: Doubleday, 1982.
[416] ———. *Stargazers and Gravediggers.* New York: Morrow, 1983.
Those articles marked with an asterisk, below, have been reprinted in reference 315.
[417] ———. *"Are the Moon's Scars Only 3000 Years Old?" *Pensée,* 1:14. Reprinted from *New York Times* (21 July 1969), early city ed.
[418] ———. *"When Was the Lunar Surface Last Molten?" *Pensée,* 1:19–21.
[419] ———. *"On Decoding Hawkins' 'Stonehenge Decoded.'" *Pensée,* 1:24–28, 50.
[420] ———. *"Is Venus' Heat Decreasing?" *Pensée,* 1:51.
[421] ———. *"H. H. Hess and My Memoranda." *Pensée,* 2:22–25.
[422] ———. *"The Orientation of the Pyramids." *Pensée,* 3:17.
[423] ———. *"Earth without a Moon." *Pensée,* 3:25.
[424] ———. *"Venus and Hydrocarbons." *Pensée,* 6:21–23.
[425] ———. *"Venus' Atmosphere." *Pensée,* 6:31–37.
[426] ———. *Harper's* (Mar. 1950), p. 18.
[427] ———. *Spectator* (27 Oct. 1950), p. 421.
[428] ———. *New Statesman and Nation* (3 Feb. 1951).
[429] ———. "Answer to My Critics." *Harper's* (June 1951), pp. 51–57.
[430] ———. "Answer to Professor Stewart." *Harper's* (June 1951), pp. 63–66.
[431] ———. "'Worlds in Collision' in the Light of Recent Finds in Archaeology, Geology, and Astronomy." Address before the Graduate College Forum of Princeton University, 14 Oct. 1953. Published as a supplement (pp. 269–301) to *Earth in Upheaval* [411].
[432] ———. "A Tempest in the Cosmos." *Book Week* (5 Sept. 1965), pp. 2, 8.
[433] ———. "Additional Examples of Correct Prognosis." Pp. 232–45 in de Grazia [78].
[434] ———, and Ralph E. Juergens. "A Rejoinder to Motz." *Yale Scientific Magazine,* 41 (Apr. 1967):14–16.
[435] ———. "Venus—A Youthful Planet." *Yale Scientific Magazine,* 41 (Apr. 1967):8–11, 32.
[436] ———. "Straka: Science or Anti-Science?" *Pensée,* 2:16.
[437] ———. "The Lion Gate at Mycenae." *Pensée,* 3:31.

[438] ———. "The Pitfalls of Radiocarbon Dating." *Pensée,* 4:12–14, 50.
[439] ———. "Astronomy and Chronology." *Pensée,* 4:38–39. Reprinted in *Peoples of the Sea* [413].
[440] ———. "Metallurgy and Chronology." *Pensée,* 5:5–9.
[441] ———. "The Velocity of Light in Relation to Moving Bodies." *Pensée,* 5:16–20.
[442] ———. "A Reply to Stiebing." *Pensée,* 6:38–42.
[443] ———. "Scarabs." *Pensée,* 6:42–45.
[444] ———. "Tiryns." *Pensée,* 6:45–46.
[445] ———. "My Challenge to Conventional Views in Science." *Pensée,* 7:10–14. Reprinted in *Velikovsky and Establishment Science* [201].
[446] ———. "The Scandal of Enkomi." *Pensée,* 10:21–23.
[447] ———. "A Concluding Retort." *Pensée,* 10:26, 49.
[448] ———. *Kronos,* 2 (no. 2) (1976):120.
[449] ———. "Quartered at Yale." *Kronos,* 2 (no. 3) (1977):49–55.
[450] ———. "Afterword." *Kronos,* 3 (no. 2) (1977):18–31.
[451] ———. "Precursors." *Kronos,* 7 (no. 1) (1981):48–54. According to a footnote, this paper was written not later than 1948.
[452] Velikovsky, Immanuel, to Harry Hess. (7 Aug. 1969), quoted in reference 350.
[453] Warshofsky, Fred. "When the Sky Rained Fire: The Velikovsky Phenomenon." *Reader's Digest* (Dec. 1975), pp. 220–40.
[454] Wassermann, Felix M. *Library Journal,* 85 (15 Mar. 1960):1114.
[455] Weinberg, Alvin M. "Science and Trans-Science." *Minerva,* 10 (1972):209–22.
[456] Weisberg, Leonard R. *Physics Today* (Sept. 1972), p. 13.
[457] Wells, Robert Dolling. *Christian Science Monitor* (15 Apr. 1950), p. 18.
[458] Westrum, Ron. "Scientists as Experts: Observations on 'Objections to Astrology.'" *Zetetic,* 1 (no. 1) (1976):34–46.
[459] ———. Letter. *Zetetic,* 1 (no. 2) (1977):107–12; with replies by Paul Kurtz and Lee Nesbit (pp. 112–17) and Lawrence E. Jerome (pp. 117–18).
[460] Whelton, Clark. "The Gordon Atwater Affair." *S.I.S. Review,* 4 (no. 4) (1980):75–76.
[461] Whiston, William. *A New Theory of the Earth.* Page references are to 2d ed. London, 1708.

[462] White, Peter. *The Past Is Human.* New York: Taplinger, 1976.
[463] Willhelm, Sidney M. *American Behavioral Scientist* (Nov. 1963), p. 23.
[464] ———. "Velikovsky's Challenge to the Scientific Establishment." *Pensée,* 3:32–35.
[465] ———. "The Velikovskian Upheaval: A Temporocentric Challenge." *Kronos,* 3 (no. 2) (1977):49–61.
[466] Winthrop, Henry. *American Behavioral Scientist* (Mar. 1964), p. 27.
[467] Witchell, Nicholas. *The Loch Ness Story.* New York: Penguin, 1975.
[468] Wolfe, Irving. "Introduction to Velikovsky." *Proceedings of Symposium,* 10–12 Jan. 1975, Montreal, pp. 1–14. Available from Saidye Bronfman Centre, 5170 Cote Saint Catherine's Road, Montreal, Quebec, Canada.
[469] Woodin, Mary Lib. *Christian Science Monitor* (15 Apr. 1950), p. 18.
[470] Wootton, Arden. *Harper's* (Aug. 1951), p. 14.
[471] Wright, Robert C. "Effects of Volatility on Rubidium-Strontium Dating." *Pensée,* 1:20. Reprinted in *Velikovsky Reconsidered* [315].
[472] *Yale Scientific Magazine.* 41 (Apr. 1967):6, 32. Unsigned biographical note about Velikovsky.
[473] York, Derek. "Lunar Rocks and Velikovsky's Claims." *Pensée,* 1:18. Reprinted in *Velikovsky Reconsidered* [315].
[474] *Zetetic Scholar.* Journal of the Center for Scientific Anomalies Research, ed. Marcello Truzzi, Department of Sociology, Eastern Michigan University, Ypsilanti, Mich. 48197.
[475] ———. Nos. 3 and 4 (1979), pp. 27–68: "A Dialogue on the Theories of I. Velikovsky: Joseph May, The Heresy of a New Synthesis." Pp. 28–47, followed by comments from David Morrison, Leroy Ellenberger, Michael Jones, Malcolm Lowery, R. G. A. Dolby, Robert McAulay, Peter J. Huber, Donald Goldsmith; epilogue by M. Truzzi.
[476] Ziman, John. *Teaching and Learning about Science and Society.* Cambridge: Cambridge University Press, 1980.

Index